Brickwork Level 2

Brickwork Level 2 has been adapted from John Hodges's classic *Brickwork for Apprentices* – the established textbook on brickwork for bricklayers. Designed to meet new requirements of the City and Guilds bricklaying programmes, this book has been written to match the latest industry-based requirements and technical developments in the field, including recent changes to the Building Regulations. Each chapter follows the syllabus and contains a section of multiple-choice questions to provide trainees with vital practice for the job knowledge and multiple-choice tests.

Highly illustrated throughout and now in full colour, this is the essential reference for qualified bricklayers and other professionals working in the construction industry, as well as students wishing to embark on a career in bricklaying.

There is also the facility to access the Support Material on the Routledge website, which includes:

- PowerPoint slides for each chapter
- Lesson plans and schemes of work
- Multiple-choice questions and answers
- Job knowledge questions and answers
- Practical drawings and mark sheets

Malcolm Thorpe is a Life Member of the Guild of Bricklayers and former Head of Construction at West Nottinghamshire College, UK. He has acted as a CITB adviser and was involved in the drafting of bricklaying programmes. He was a verifier for CITB and spent several years producing examination material for examining bodies. He also spent over thirty years designing and organising Guild of Bricklayers local and national bricklaying competitions.

Brickwork Level 2

For Construction Diploma, Technical Certificate and Apprenticeship Programmes

Second Edition

Malcolm Thorpe

Routledge
Taylor & Francis Group

LONDON AND NEW YORK

Second edition published 2021
by Routledge
2 Park Square, Milton Park, Abingdon, Oxon, OX14 4RN

and by Routledge
52 Vanderbilt Avenue, New York, NY 10017

Routledge is an imprint of the Taylor & Francis Group, an informa business

First edition published by Routledge 2010

British Library Cataloguing-in-Publication Data
A catalogue record for this book is available from the British Library

Library of Congress Cataloging-in-Publication Data
A catalog record has been requested for this book

ISBN: 978-0-367-62549-8 (hbk)
ISBN: 978-0-367-62536-8 (pbk)
ISBN: 978-1-003-10965-5 (ebk)

Typeset in Utopia
by MPS Limited, Dehradun

Access the Support Material: www.routledge.com/9780367625368

Contents

Preface . . . . *vi*

Chapter 1   The Construction Industry (Community Buildings) 1

Chapter 2   Health and Safety in the Construction Industry 17

Chapter 3   Communication and Records 35

Chapter 4   Construction Technology 95

Chapter 5   Moving and Handling Resources 155

Chapter 6   Solid Walls and Piers 199

Chapter 7   Constructing Cavity Walls with Openings 263

Chapter 8   Cladding to Timber, Steel and Concrete Frames 311

Chapter 9   Thin Joint Masonry 339

Chapter 10 Bridging Openings with Lintels and Arches 359

Chapter 11 Domestic Drainage 385

Chapter 12 Answers to Multiple-Choice Questions 425

Index 429

Preface

Changes in construction training have led to the need to update this series of books which incorporate the new Occupational Standards.

The content of each book follows the On-Programme Training Specification.

The On-Programme specification is divided into Knowledge Requirements (K) and Skill Requirements (S).

After the initial chapter, which gives the apprentice an insight into the construction industry they are entering, each chapter follows very closely the Training Specification for Technical Certificate, Apprenticeship and Diploma units.

The aim of each book is to provide an informative resource and workbook for all building craft apprentices. It can be used to provide teaching and assessment material, or used simply to reinforce college lectures.

Each chapter has a set of multiple-choice questions designed to test your level of knowledge before you move on to the next chapter.

There is also the facility to access the Support Material on the Routledge website, which includes lesson plans, schemes of work, completed units tracking sheets, mapping sheeting, PowerPoint displays, activity packs, job knowledge questions and answers, multiple-choice questions and answers, practical training units and practical competency units with marking sheets.

It can be accessed here: www.routledge.com/9780367625368

Malcolm Thorpe

CHAPTER *1*

The Construction Industry (Community Buildings)

This chapter will cover the following City and Guilds units:

- Apprenticeship Level 2   Module 1   K1.1, K1.2
- Certificate Level 2   Unit 201   1.1, 1.2, 1.3
                                                    2.1, 2.2
- Diploma Level 2   No comparable unit

This chapter is about:

- The construction industry
- Building types
- Types of community buildings
- Restoration and heritage
- Types of construction

The following Skills and Knowledge requirements will be covered:

- Know building types
- Types of construction

The following Certificate learning outcomes will be covered:

- Understand how to work in the construction industry

The following Diploma learning outcomes will be covered:

- This chapter has no comparable Diploma units but gives the student an early introduction to the construction industry

Introduction

When students are thinking of entering the building trade they may ask many questions. The three main questions are:

- What is the construction industry?
- What can the construction industry offer me?
- What type of education will I need?

The construction industry

The construction industry is one of the largest employers of labour in the country, with a labour force of just over one million, a figure that has dropped steadily over the past years.

Construction means creating, not only the houses we need to live in but many other buildings such as schools, hospitals and shopping centres.

The majority of buildings and structures are designed and constructed for a specific purpose. The use of the building will determine the size, shape, style and ultimately the cost.

Every person employed in the construction industry makes a direct contribution to the community in general but also to the nation.

The industry is made up of a large number of firms which can be classified as:

- builders
- contractors
- subcontractors, etc.

The firms range in size from one-person firms to multinational companies.

- A small company is defined as having between one and 49 employees.
- A medium company is defined as having between 50 and 249 employees.
- A large company is defined as having more than 250 employees.

There are also several different types of construction work to consider when thinking of joining the construction industry.

The whole industry can be further divided into four:

- *New work* refers to all types of building work and services which are about to start.
- *Maintenance work* refers to any work on an existing building which requires damaged or out-of-date items to be brought up to an acceptable standard.

Examples of maintenance work include new kitchen and bathroom units, or external brickwork requiring repointing, etc.

- *Refurbishment* is when an old building has been allowed to fall into a state of disrepair and it needs to be brought back to standard. Changing existing buildings for another purpose is also classed as refurbishment.

An example is when an old warehouse has been changed into a block of flats.

- *Restoration work* is when an old building is brought back to its original state.

Examples are old historic buildings bought by the National Trust and then painstakingly restored to their former glory.

The construction process

The construction process is said to be the most complex of all industries.

People employed in manufacturing industries travel each day to the same place and do the same type of work.

In the construction industry the employees move to a different place of work as soon as the particular job has been completed. The distance depends on the nature and size of the contractor and the availability of work.

No two construction sites are ever the same, and it is seldom that more than a few dwellings are the same.

The construction industry does not lend itself to production-line methods, so it is very labour intensive.

The construction team is therefore comprised of people possessing a vast range of skills, from the tradespeople to the professionals.

The construction industry differs from other industries in the following ways:

- Work is carried out in the open and is subject to stoppages from the weather.
- Every day the plan of work is different.
- The labour force is not static and can change daily.
- Great distances have to be travelled by employees, so they are often many miles from the head office.
- Every job is different, so there is no repetition through which employees can produce more after gaining experience.
- Many of the contracts are completed by one person after being designed by another person.
- Safety in the industry has a very poor record.
- The industry is very labour intensive.

All the above statements can cause many problems and it is very difficult for any one person to rectify them; therefore the construction team becomes very important indeed.

Types of building

Many different types of construction are required to fulfil the needs of today's ever-demanding society.

These consist of the following and are shown in Figure 1.1:

- dwelling units – for people to live in
- communal buildings – for all people to share
- industrial units – for people to work in
- recreational units – for people to relax in
- communications – roads, rail, sea and air networks – to allow people to move from one place to another.

Community buildings

For the purpose of Level 2, only buildings for communal purposes will be dealt with.

These buildings cover those found in recreational, educational, spiritual and leisure sectors. Buildings in this category are used for the benefit of the community. They are used to accommodate the facilities that society provides for:

- health care
- welfare
- education
- entertainment.

Examples of buildings that provide for the spiritual needs of the community are:

- churches
- chapels
- cathedrals.

Examples of community buildings include:

- shopping centres
- hospitals
- police and fire stations.

Other buildings in this category provide buildings for leisure, such as:

- public houses
- dance halls
- theatres
- cinemas
- concert halls.

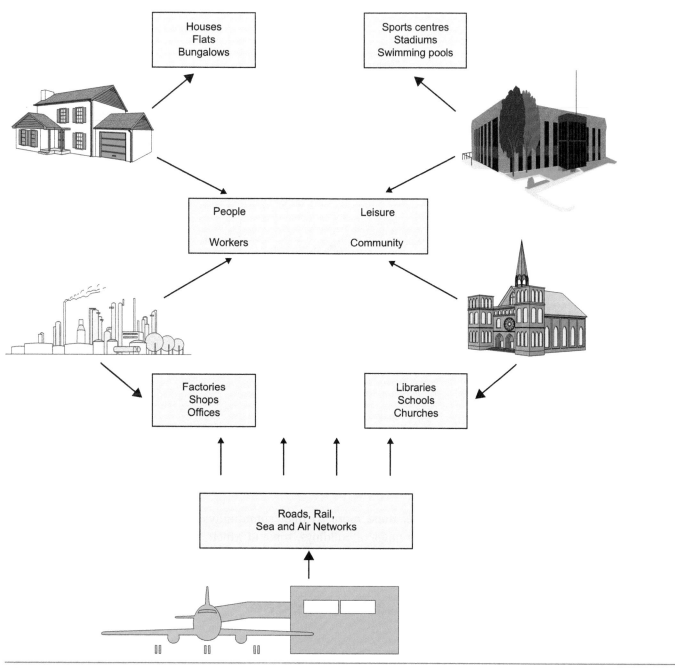

FIGURE 1.1
Types of construction

Buildings designed and constructed for education include:

- schools
- colleges
- libraries.

Buildings in this category are designed and built for the use of all the community. The list is by no means exhaustive.

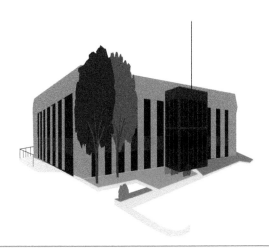

FIGURE 1.2
Communal buildings

Classification of communal buildings

All buildings are designed to fulfil a role, none more so than community buildings. This category covers a very wide range of buildings, some of which are shown in Figure 1.2.

Classification of buildings by purpose or use

The classes of use for England as set out in the Town and Country Planning (Use Classes) Order 1987 (as amended) puts uses of land and buildings into various categories known as 'Use Classes'. The relevant building regulations also need to be considered for any proposed change of use.

Whenever you are thinking of creating, changing or demolishing buildings or land it is important to understand what use class they fall under. This also applies if you are changing how a building or land is used (for instance, turning a residential property into a business) but not necessarily making physical changes.

Sometimes, depending on which use class they come under, work carried out on buildings or land may be covered by permitted development rights. This means that you do not need to apply for planning permission but you may still need to seek prior approval or notify us before you make the changes.

List of use classes

Part A

A1 Shops

Shops, retail warehouses, hairdressers, undertakers, travel and ticket agencies, post offices, pet shops, sandwich bars, showrooms, domestic hire shops, dry cleaners, funeral directors and internet cafes

A2 Financial and professional services

Financial services such as banks and building societies, professional services (other than health and medical services) and including estate and employment agencies. It does not include betting offices or pay day loan shops – these are now classed as "sui generis" uses (see below)

A3 Restaurants and cafés

For the sale of food and drink for consumption on the premises – restaurants, snack bars and cafes

A4 Drinking establishments

Public houses, wine bars or other drinking establishments (but not night clubs) including drinking establishments with expanded food provision

A5 Hot food takeaways

For the sale of hot food for consumption off the premises.

Part B

B1 Business

Uses which can be carried out in a residential area without detriment to its amenity. This class is formed of three parts:

- B1(a) Offices - Other than a use within Class A2 (see above)
- B1(b) Research and development of products or processes
- B1(c) Industrial processes

B2 General industrial

Use for industrial process other than one falling within class B1 (excluding incineration purposes, chemical treatment or landfill or hazardous waste)

B8 Storage or distribution

This class includes open air storage.

Part C

C1 Hotels

Hotels, boarding and guest houses where no significant element of care is provided (excludes hostels)

C2 Residential institutions

Residential care homes, hospitals, nursing homes, boarding schools, residential colleges and training centres

C2A Secure residential institution

Use for a provision of secure residential accommodation, including use as a prison, young offenders institution, detention centre, secure training centre, custody centre, short term holding centre, secure hospital, secure local authority accommodation or use as a military barracks

C3 Dwelling houses – This class is formed of three parts

C3(a)

covers use by a single person or a family (a couple whether married or not, a person related to one another with members of the family of one of the couple to be treated as members of the family of the other), an employer and certain domestic employees (such as an au pair, nanny, nurse, governess, servant, chauffeur, gardener, secretary and personal assistant), a carer and the person receiving the care and a foster parent and foster child

C3(b)

covers up to six people living together as a single household and receiving care e.g. supported housing schemes such as those for people with learning disabilities or mental health problems

C3(c)

allows for groups of people (up to six) living together as a single household. This allows for those groupings that do not fall within the C4 HMO definition, but which fell within the previous C3 use class, to be provided for i.e. a small religious community may fall into this section as could a homeowner who is living with a lodger

C4 Houses in multiple occupation

Small shared houses occupied by between three and six unrelated individuals, as their only or main residence, who share basic amenities such as a kitchen or bathroom.

Part D

D1 Non-residential institutions

Clinics, health centres, crèches, day nurseries, day centres, schools, art galleries (other than for sale or hire), museums, libraries, halls, places of worship, church halls, law court. Nonresidential education and training centres

D2 Assembly and leisure

Cinemas, music and concert halls, bingo and dance halls (but not night clubs), swimming baths, skating rinks, gymnasiums or area for indoor or outdoor sports and recreations (except for motor sports, or where firearms are used).

Sui generis

'Sui generis' is a Latin term that, in this context, means 'in a class of its own'.

Certain uses are specifically excluded from classification by legislation, and therefore become 'sui generis'. These are:

- theatres

- amusement arcades/centres or funfairs

- launderettes

- fuel stations

- hiring, selling and/or displaying motor vehicles

- taxi businesses

- scrap yards, or a yard for the storage/distribution of minerals and/or the breaking of motor vehicles

- 'Alkali work' (any work registerable under the Alkali, etc. Works Regulation Act 1906 (as amended))

- hostels (providing no significant element of care)

- waste disposal installations for the incineration, chemical treatment or landfill of hazardous waste

- retail warehouse clubs

- nightclubs

- casinos

- betting offices/shops

- pay day loan shops

Legislation

As the population has steadily grown from the Middle Ages through until the modern day the standard of dwellings has improved.

Man has created some form of dwelling from very simple shelters to protect himself from the weather and wild animals until the present elaborate designs planned for modern day comfort.

The design and construction of buildings has improved to include the advances in technology.

It took the Great Fire of London to make the Government realise that some form of control was required.

The London Building Act, controlling the rebuilding of the City of London, came into force in 1667. The main aim of this Act was 'fire protection'.

Various other Acts followed including the Factories Act, The Public Health Act, The Building Byelaws, The Town and Country Planning Act.

In 1965 the Building Regulations were enacted. This was the first set of regulations covering buildings to apply nationally.

The Building Regulations are designed to set minimum standards for all building work and to safeguard public health and safety. They are administered by the local authorities through their building surveyors department.

Heritage

Types of heritage

There are three types of sites: cultural, natural, and mixed.

Cultural heritage sites include hundreds of historic buildings and town sites.

Natural heritage sites include important archaeological sites.

Mixed heritage includes works of monumental sculpture or painting.

Typically, a heritage building means a structure that requires preservation because of its historical, architectural, cultural, aesthetic or ecological value.

Archaeological sites, ruins and remains are also covered under the term heritage sites.

Historic preservation is an endeavour that seeks to preserve, conserve and protect buildings, objects, landscapes or other artefacts of historical significance.

As we move into the future with the explosion of new technologies, historic preservation provides an anchor to our past.

Restoration, refurbishment, preservation and renovation

Building restoration

Restoration is the act of repairing or renewing something.

An example of restoration is fixing an old house to its original state as seen in Figure 1.3.

Restoration work is most commonly undertaken on historic buildings; accurately recreating its form, features and character as it appeared at a particular time, while protecting its heritage value.

Building refurbishment

Refurbishment is a term used to describe a process of improvement by cleaning, decorating and re-equipping. It may also include elements of retrofitting with the aim of making a building more energy efficient and sustainable. See Figure 1.4.

The lifecycle of a building can be significantly extended by effective refurbishment.

FIGURE 1.3
Restored building

FIGURE 1.4
Refurbished building

Building preservation

Preserving historic buildings is crucial to retaining our nation's heritage and history. It is also environmental and practical, basically large scale recycling. It reduces the need for new materials and the other costs of demolition.

FIGURE 1.5
Preserved building

Preserving and restoring the old buildings is also important because those old monuments are the reflection of our history, they help us to understand and respect people who lived in different eras with different habits and traditions. See Figure 1.5.

FIGURE 1.6
Renovated building

Building renovation

If you restore an old 1950s house, you use period materials and effects. Restoration always means a putting back.

A renovation of the same property might include new things such as a new bathroom and kitchen. A typical renovation project is shown in Figure 1.6.

Multiple-choice questions

Self-assessment

This section of the book is designed to allow you to check your level of knowledge. The section consists of revision questions for this chapter. The questions are all multiple choice and have four possible answers. The answers are to be found at the end of the book.

The main type of multiple-choice question will be the four-option multiple-choice question. This will consist of a question or statement, known as the stem, followed by a choice of four different answers, called the responses. Only one of these responses is the correct answer; the others are incorrect and are known as distracters.

You should attempt to answer the questions by choosing either (a), (b), (c) or (d).

Example

The person employed by the local authority to ensure that the Building Regulations are observed is called the:

    (a) clerk of works

    (b) building control officer

    (c) council inspector

    (d) safety officer

The correct answer is the building control officer, and therefore (b) would be the correct response.

The construction industry

Question 1 Classification of buildings is set out in the:

    (a) Town and country Planning Order

    (b) The Factories Act

    (c) The Public Health Act

    (d) The Building Bylaws

Question 2 State the type of community unit shown.

    (a) library

    (b) school

    (c) church

    (d) theatre

Question 3 Groups of shops all under one roof are known as:
- (a) high-street shops
- (b) shopping centres
- (c) community centres
- (d) arenas

Question 4 The act of repairing or renewing a building is known as:
- (a) conserving
- (b) preserving
- (c) renewing
- (d) restoring

Question 5 Archaeological sites are known as:
- (a) brown-field sites
- (b) heritage sites
- (c) historical sites
- (d) building sites

Question 6 Which of the following communal buildings has been designed for educational needs?
- (a) church
- (b) hospital
- (c) school
- (d) football stadium

Question 7 Identify the following communal building:

- (a) Theatre
- (b) Hospital
- (c) Public house
- (d) School

Question 8 A person who designs a communal building is known as the:
- (a) architect
- (b) surveyor
- (c) estimator
- (d) client

CHAPTER 2

Health and Safety in the Construction Industry

This chapter will cover the following City and Guilds units:

- Apprenticeship Level 2   Module 2   K2.1, K2.2, K2.3, K3.1, K3.2, K3.3, K3.4
                               Module 10   S3.1, S3.2, S3.3
                               Module 9   S1.1, S2.1
- Certificate Level 2   Unit 201   2.4
- Diploma Level 2   Unit 201/601   1.1, 1.2, 1.3, 1.4, 1.5, 2.1, 2.2, 2.3, 2.4, 2.5, 2.6, 3.1, 3.2, 3.3, 3.4, 3.5, 4.1, 4.2, 4.3, 6.1, 6.2, 6.3, 6.4, 7.1, 7.2, 7.3, 7.4, 8.1, 8.2, 8.3, 8.4, 8.5, 9.1, 9.2, 9.3, 9.4

This chapter is about:

- Awareness or relevant current statutory requirements and official guidance
- Personal responsibilities relating to workplace safety, wearing appropriate protective equipment and compliance with warning signs
- Personal behaviour in the workplace
- Security in the workplace
- Relationships

The following Skills and Knowledge requirements will be covered:

- Understand how site procedures contribute to maintaining health and safety
- Maintain a safe and secure work environment for customers during construction works

The following Certificate learning outcomes will be covered:

- Understand Health and Safety guidance used during the construction process

The following Diploma learning outcomes will be covered:

- Know health and safety regulations, roles and responsibilities
- Know accident and emergency reporting procedures
- Know how to identify hazards in the workplace
- Know about health and welfare in the workplace
- Know how to handle materials and equipment safely
- Know about access equipment and working at heights
- Know how to work with electrical equipment in the workplace
- Know how to use PPE
- Know the cause of fire and emergency procedures

Safety legislation

Safety in the workplace will be covered only briefly in this chapter as it has already been covered in depth in Chapter 2 of Level 1. If more detailed information is required please refer to Level 1.

The construction industry is often involved in very difficult and often hazardous sites. It is therefore very important that the new recruit is aware of these dangers and that there are various regulations in place to control and reduce these possible hazards.

Prevention of hazards in the workplace

Hazards within a workplace can occur because of several circumstances. There may be faults in equipment, tools, stored substances, dangerously stacked materials, materials obstructing safe access, or simply a lack of site safety.

The health and safety of employees at their workplace and any other persons at risk through work activities are covered through various Acts of legislation and regulations.

> **NOTE**
>
> As Regulations are updated regularly it is always advisable to check with the latest version.

These include the following:

- The Health and Safety at Work Act 1974

- The Control of Substances Hazardous to Health Regulations 2002 (COSHH)

- The Noise at Work Regulations 2005

- Work at Height Regulations 2005

- Reporting of Injuries, Diseases and Dangerous Occurrences Regulations 2013 (RIDDOR)

- The Personal Protective Equipment at Work Regulations amended 2018

- The Fire Precautions (Workplace) Regulations amended 1999

- Provision of the Use of Work Equipment Regulations 1998 (PUWER)

- The Electricity at Work Regulations 1989

- The Manual Handling Operations Regulations 1992

- Control of Vibration at Work Regulations

- Lifting Operations and Lifting Equipment Regulations 1998 (LOLER)

- The Construction (Design and Management) Regulations 2015

The main health and safety legislation applicable to building sites and workshops is covered by the Health and Safety at Work Act 1974.

Health and safety at work ACT 1974

The four main objectives of the HASAWA are:

- To secure the health, safety and welfare of all persons at work.

- To protect the general public from risks to health and safety arising from out of work activities.

- To control the use, handling, storage and transportation of explosives and highly flammable substances.

- To control the release of noxious or offensive substances into the atmosphere.

This Act requires employers to ensure so far as is reasonably practicable the health and safety of their employees, other people at work and members of the public who may be affected by their work.

Employees have to co-operate with their employer on health and safety matters and not do anything that puts them or others at risk.

Employees should be trained and clearly instructed in their duties.

The main purpose of this Act is to cover all aspects of safety.

The framework promotes, stimulates and encourages high standards of health and safety in the workplace.

The Act involves everyone, management, employees, the self-employed, the employees' representatives, the controllers of premises, and the manufacturers of plant, equipment and materials, in matters of health and safety.

The Act also deals with the protection of the public, where they may be affected by the activities of people at work.

Outline of the Act

The Act itself is very complex and is an extensive document with numerous parts and sections. There are four main parts.

Part 1 of the Act describes:

Employers' and management duties:

1. Provide and maintain a safe working environment.

2. Ensure safe access to and from the workplace.

3. Provide and maintain safe machines, equipment and methods of work.

4. Ensure the safe handling, transport and storage of all machinery, equipment and materials.

5. Provide their employees with the necessary information, instruction, training and supervision to ensure safe working.

6. Prepare, issue to employees and update as required a written statement of the firm's safety policy.

7. Involve trade union safety representatives (where appointed) with all matters concerning the development, promotion and maintenance of health and safety requirements.

Employees' duties:

1. Take care at all times and ensure that they do not put themselves, their workmates or any other person at risk by their actions.

2. Co-operate with their employers to enable them to fulfil the employer's health and safety duties.

3. Use the equipment and safeguards provided by the employers.

4. Never misuse or interfere with anything provided for health and safety.

Safety procedures and documentation

In order to comply with the various safety legislation, an employer is required:

- To display notices and certificates – e.g. a copy of a valid fire certificate.

- To notify relevant records – e.g. commencement of any building works likely to last in excess of six weeks has to be notified to the relevant authority.

- To keep relevant records –e.g. the accident book in which details of *all* accidents are recorded.

Workplace safety

Accidents

When joining the construction industry it is important to remember that you will be joining an industry with one of the highest injury and accident ratings. It therefore cannot be stressed enough that you could be at constant risk unless you start as you intend to continue, with a good safety attitude.

Any type of work carried out by the construction industry is often difficult and hazardous. Every site will be different, and therefore every site will bring possible new dangers.

It is of utmost importance that *all* trainees are capable of using hand tools and equipment efficiently and safely at an early stage in their development.

Furthermore, they should be aware of the causes of accidents and be able to take action and deal with any accident that may occur.

An accident is an unexpected or unplanned happening which results in personal injury or damage, sometimes death.

Reported accidents are those which result in death, major injury and more than three days' absence from work or are caused by dangerous occurrences reported to the HSE.

Every day a large number of the industrial accidents that are reported involve construction workers.

ACCIDENTS DO NOT JUST HAPPEN, THEY ARE CAUSED.

Learning to spot a dangerous situation is not as difficult as it sounds, because accidents follow a regular pattern. The same kind of accident happens over and over again. Every day of the year, all over the country, the same set of dangerous conditions builds up and the same unsafe acts take place.

Do any of the things you normally see and do at work add up to a source of danger? Next time you are tempted to take a risk – STOP and THINK again.

Types of hazard

Everyone involved in the construction industry should be aware of the possible dangers and hazards on the construction site.

Site safety will be improved if everyone is safety conscious.

Types of hazard include the following:

- falling objects
- falls of operatives
- transportation of plant and materials
- electricity
- machinery and equipment
- fire and explosions.

Personal protective equipment

Depending on the type of workshop or site situation, the wearing of correct safety clothing and safe working practices are the best methods of avoiding accidents or injury. On some sites certain PPE is compulsory.

All construction operatives have a responsibility to safeguard themselves and others. Making provision to protect oneself often means wearing the correct protective clothing and safety equipment.

Your employer is obliged by law to provide PPE, a selection of which is shown in Figure 2.1.

Safety signs

As you go about your work on the building or construction site you will see various signs and notices. Your employer will give you instruction on what they mean and what you should do when you see one.

Safety signs fall into four separate categories, which can be recognized by their shape and colour. Sometimes they may be just a symbol; others may include letters or figures and provide extra information such as the clearance height of an obstacle or the safe working load of a crane.

Remember

It is your responsibility to act in a safe manner.

FIGURE 2.1
A selection of personal protective equipment

The four categories are:

Prohibition signs

- Shape – circular
- colour – red border and cross bar; black symbol on white background
- meaning – shows what must not be done
- example – no smoking.

Mandatory signs

- Shape – circular
- colour – white symbol on blue background
- meaning – shows what must be done
- example – wear hand protection.

Warning signs

- Shape – triangular
- colour – yellow background with black border and symbol
- meaning – warns of hazard or danger
- example – caution, fork-lift truck working.

Information signs

- Shape – square or oblong
- colour – white symbols on green background
- meaning – indicates or gives information of safety provision
- example – first aid point.

Signs with supplementary text

Any of the symbols shown on the last page may also contain text. A few examples of these are shown in Figure 2.2.

Signs may be produced and erected according to the situation, as shown in Figure 2.3. For example, where there is a dangerous hazard a sign can be produced and placed in a prominent position to warn anyone approaching to take care.

(a) **Prohibition Sign**

(b) **Mandatory Sign**

No parking

Protective footwear must be worn

(c) **Warning Sign**

(d)

Danger
Men at work

Information Sign

NO TOOLS OR
EQUIPMENT ARE
STORED OVERNIGHT
IN THIS VEHICLE

FIGURE 2.2
A selection of signs: (a) prohibition sign; (b) mandatory sign; (c) warning sign; (d) information sign

Security arrangements

It is the responsibility of everyone on the work site to ensure that the security of that site is maintained. Security can take many forms and they are all equally important

Visual security

- Alarms – positioned in an accessible place within view of the general public

- bars, mesh and locks – fitted to glass-panelled doors and windows

- padlocks, padlock and chains – fitted to compound gates, pieces of plant and machinery

- lighting – floodlights and movement-activated lights

- security firms.

Individual security

It is the responsibility of all employees to contribute to the overall security of the firm, for example:

- Tidiness – Do not invite crime by leaving tools and equipment where they may be easily seen. If there is a secure store, lock them away.

- Plant – If possible, return all plant to a secure compound, or if necessary immobilize.

- Unauthorized access – It is the responsibility of all employees on site to challenge anyone who they feel has no authorization to be within a particular area. (Politeness is the best approach.)

If, despite the security measures taken, your site is breached, there are certain procedures you should follow. These should be given to you by your line manager, and may include:

- reporting the incident to the site supervisor

- reporting the incident to the police

- checking the inventory to find out what has been taken

- recording damage done to the premises and/or equipment.

Remember

Vandalism and theft do not always happen at night.

| SCAFFOLDING INCOMPLETE | THIS IS A HARD HAT AREA | CAUTION Unsafe Do not use Signed: |

FIGURE 2.3
Types of notice

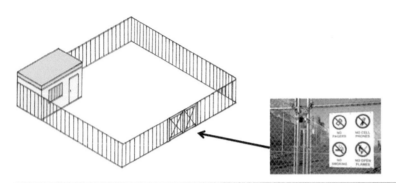

FIGURE 2.4
Site compound with cabin and secure gates

Site security

Always take care of tools and equipment you are working with and report any suspicious persons or events to your line manager.

Site access

It is important on larger sites to have a signing in and signing out procedure to increase the security of the site and its workers.

On larger sites all employees are checked/searched to make sure they are fit to work. They are provided with facilities for parking their cars, changing into work clothes, rest areas for eating food and washing and toilet facilities.

Personal security

All PPEs provided must be worn at all times in the work areas.

Some sites will not allow newspapers, food or drink in the work areas and the provided toilet areas have to be used.

Some sites use a card system similar to football referees, a yellow card is handed out for an offence.

Three yellow cards and a red is issued which could lead to refusal to enter the work area.

Loss of goods and materials

The contractor has a duty to prevent the entry to the work area by providing an adequate security fence.

Lockable cabins and storage areas should also be provided to protect materials and equipment.

The whole site should be surrounded with a security fence and adequate security to the gates and a secure site cabin as shown in Figure 2.4.

Emergency procedures

Responding to emergencies

It is important that all new trainees are aware of the emergency procedures, which could change according to the type of site or workplace they are in.

From day one you should be aware of what to do in the event of a fire or an accident. Should an emergency happen you should be able to:

- Know what to do – acting quickly and calmly, carry out the correct procedure

- Follow the fire procedure – take the correct action in the event of discovering a fire:

  1. Select and use the correct type of fire extinguisher. (Only if the fire is small enough for you to put it out.)

  2. Call for help, sound the alarm.

  3. Telephone the Fire Service: 999. Give the correct address of the building.

  4. Leave the building by the nearest exit.

  5. Go directly to the assembly area. Await the roll call.

Accidents

An accident involving injury to a person can happen at any time. It may be a workmate who has fallen off a ladder or someone with a burn or a cut, or who has fainted. To help them when they most need it you should know what to do!

Immediate action

1. Unless you are a fully trained first aider – *do not attempt to treat the injured person*. (Only move an injured person if their life is in danger, e.g. danger from fire.)

2. GET HELP. Report the accident to a person in charge or:

3. Telephone the Emergency Services: 999.
   When your call is answered you should have the following information at hand:

   - type or types of services required – fire, ambulance, police

   - type of accident

   - location/address at which it has happened

   - telephone number you are calling from

   - your name.

Remember

An accident is an event causing injury or damage that could have been avoided by following correct methods and procedures.

Risk assessments

The employer has a duty to protect the workforce as far as is reasonably practicable. Risk assessment is a very important part of protecting all site operatives.

Risk assessment is simply a careful examination of what could cause harm to the workforce. The workforce has a right to be protected from harm caused by a failure to take reasonable control measures.

In the construction industry risk assessments are carried out by experienced people who have been taught to identify what risks are possible when carrying out tasks.

There are five main steps to risk assessments:

- Step 1 – Identify the hazard.
- Step 2 – Decide who might be at risk and how.
- Step 3 – Evaluate the risks and decide on the best precautions.
- Step 4 – Record your findings and implement them.
- Step 5 – Review your assessment and update if necessary.

Site organisation

Welfare facilities

Certain minimum safety, health and welfare considerations are laid down in relation to construction sites by the Factories Acts and Working Rule Agreements, these call for:

1. Canteens (places to eat food)
2. Drying and changing rooms
3. Shelter from inclement weather
4. Washing and toilet facilities

Portable welfare facilities can be provided as shown in Figure 2.5.

It may also be possible when refurbishing existing premises to use the facilities already in existence.

Table 2.1 shows a chart containing the facilities to be provided according to the number of workers on site.

Whenever possible toilets should be flushed by water, but if not possible, use chemical toilets.

Rooms containing sanitary conveniences should be adequately ventilated and lit.

Men and women may use the same toilet, provided it is a lockable room and is suitably positioned away from any urinals which may also been provided.

A wash hand basin with water, soap and towels or dryers should be close to the toilets if the toilets are not near other washing facilities provided on the site.

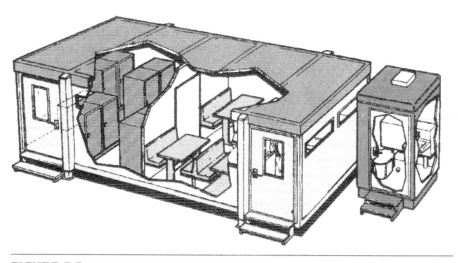

FIGURE 2.5
Typical pre-formed cabins for all requirements

Table 2.1 Welfare facilities

WELFARE FACILITIES

No of persons employed by the contractor on site		0 5 10	20 25	40	50	100
First Aid Box	Box to be clearly marked with named person		First aid boxes	First aid boxes and trained person		
Stretcher Ambulance				Stretcher provided. Local Authority informed.		
First aid Room	To be used only for treatment and in charge of trained person			Where persons on site exceed 250, each employer of more than 40 must provide First aid room		
Shelter and clothing	All persons to have shelter and place for depositing clothing	(Where possible)		Provision for warming persons and drying wet clothes		
Meals room	All persons to have drinking water provided and facilities for boiling water and eating meals			Facilities for heating food if hot meals are not available on site		
Washing facilities	All persons on site for more than 4 hours to have washing facilities		When the site lasts over 6 weeks Hot and cold water, soap and towels		More than 1 year 4 wash places per 35.	
Sanitary facilities	To be maintained and kept clean – provision for lighting		1 convenience for every 25 persons		1 for every 35	

NOTES: Washing facilities to be close to meals room
Protective clothing to be provided where person is required to work in inclement weather
Sub-contractors may use the facilities provided by another contractor and their work force is include in the total for the site

Health and hygiene

Certain precautions must be taken to ensure that the health of employees in construction firms is protected against hazards, as mentioned in the previous section. As far as is practicable, their health must also be protected.

Vulnerable parts of the body

The health of the site operatives can be divided into the following areas of the human body:

- Skin – one of the most common problems with the skin is dermatitis. This is caused by contact between the skin and the many cements and plasters on site. To reduce the problem barrier creams could be used or appropriate gloves worn.

- Eyes – protection of the eyes has been mentioned previously, but as they are the only ones you have it is important to take extra care and use the appropriate glasses or goggles for the job.

- Ears – again, most sites provide a selection of ear protectors or plugs to be used when working with or close to noise.

- Lungs – many construction operations involve dust. It is therefore very important to protect yourself against inhaling any harmful dust. Protective breathing apparatus or simple disposable masks should be available.

Personal hygiene

Always keep yourself clean and tidy; just because you are in one of the dirtiest occupations, there is no need to look untidy.

If you are working in a client's home you need to present yourself correctly. Always wash regularly and have your work clothes regularly washed.

Wash your hands after going to the toilet and before eating and drinking.

Site control

Site layout must also allow for adequate management and control facilities.

There are FOUR aspects of quality control which will have to be catered for on the construction site.

1/2. Materials delivered to the site should be checked for correct amount and quality.

3. All components delivered or components manufactured on the site will have to be inspected to ensure they are according to the specification provided.

4. Workmanship.

Remember

Do not overcomplicate the process. If you run a small business and you are confident you understand what is involved, you could do the assessment yourself. You do not have to be a health and safety expert.

Most of the risks come from tripping, slipping and moving heavy loads.

If you are an employer of a large company you should ask a health and safety advisor to help out.

Remember

A hazard is anything that can cause harm, such as chemicals, electricity or working from ladders.

The risk is the chance, high or low, that an employee could be harmed by these and other hazards.

FIGURE 2.6
Multi storey offices

Office facilities

Office and welfare facilities must be provided for the site management staff

Bills of Quantities also often require accommodation for the client's representatives.

Sound, well heated and well-lit offices are a sound investment.

These are normally portable and frequently modular which gives greater versatility.

The size and quantity of offices varies from contract to contract.

The minimum requirements of such accommodation are governed by the Offices, Shops and Railway Premises Act 1963, unless they are:

1. Mobile units in use for not more than 6 months.

2. Fixed units in use for not more than 6 weeks.

3. Any type of unit in use for not more than 21 hours per week.

4. Office for exclusive use of self-employed personnel.

5. Office used by family only staff.

Site accommodation

Site accommodation could be temporary on open sites but could be part of the structure on a refurbishment contract.

Multiple-choice questions

Self-assessment

This section of the book is designed to allow you to check your level of knowledge. The section consists of revision questions for this chapter. The questions are all multiple choice and have four possible answers. The answers are to be found at the end of the book.

The main type of multiple-choice question will be the four-option multiple-choice question. This will consist of a question or statement, known as the stem, followed by a choice of four different answers, called the responses. Only one of these responses is the correct answer; the others are incorrect and are known as distracters.

You should attempt to answer the questions by choosing either (a), (b), (c) or (d).

Example

The person employed by the local authority to ensure that the Building Regulations are observed is called the:

    (a) clerk of works

    (b) building control officer

    (c) council inspector

    (d) safety officer

Health and safety in the construction industry

Question 1 The official body that enforces health and safety is:

    (a) the local authority

    (b) the public health authority

    (c) the Health and Safety Executive

    (d) the employer

Question 2 When should you wear safety goggles on site?

    (a) at all times

    (b) in strong sunlight

    (c) when cutting bricks

    (d) when told to

Question 3 What word is missing from the sign shown?

    (a) No parking

    (b) No smoking

    (c) Danger men at work

    (d) Hard hat area

Question 4 What should you do when you hear the fire alarm ringing?

    (a) leave the site

    (b) go to the fire assembly point

    (c) go to the canteen

    (d) go to see the fire

Question 5 What shape should an information sign be?

    (a) circular

    (b) oval

    (c) square or oblong

    (d) triangular

Question 6 Which of the following current regulations covers working in dust?

    (a) Health and Safety at Work Act

    (b) Control of Substances Hazardous to Health

    (c) Work at Height Regulations

    (d) Personal Protective Equipment Regulations

Question 7 When should you wear a hard hat on site?

    (a) When you are building a boundary wall

    (b) When it is raining

    (c) When there is a danger of objects falling from scaffolds or when in a trench

    (d) When you are pointing a wall

Question 8 What should you do before lifting a heavy load?

    (a) put on gloves

    (b) bend your knees

    (c) assess the weight to be lifted

    (d) keep your back straight

CHAPTER 3

Communication and Records

This chapter will cover the following City and Guilds units:

- Apprenticeship Module 3   K1.1, K3.1, K3.5, K5.1, K5.2
                  Module 10   S2.1, S4.1, S4.2
- Certificate   Unit 201   1.3, 2.2, 2.3, 3.2
                  Unit 303   1.1, 1.2
                  Unit 305   1.3
- Diploma   Unit 202/602 10.1, 10.2, 10.3, 16.1, 16.2, 16.3, 16.4, 16.5, 16.6, 16.7
                  Unit 204   1.1, 1.2, 1.3, 2.1, 2.2, 2.3, 2.8
                  Unit 205   1.1, 1.2, 1.3, 1.4, 1.5, 1.6, 1.7, 2.1, 2.2, 2.3, 2.4, 2.5, 2.6, 2.6,
                                3.1, 3.2, 3.3, 3.4, 4.5, 4.1, 1.2, 1.3, 4.1, 4.5, 5.1, 5.2, 5.3, 5.4,
                                6.1, 6.2, 6.3, 6.4, 6.5, 7.1, 7.2, 7.3, 7.1, 7.5, 7.6, 7.7, 7.8, 7.9,
                                8.1, 8.2, 8.3, 8.4, 8.5, 8.6, 8.7, 8.8, 8.9
                  Unit 206   1.2, 1.2, 1.3, 1.4

This chapter is about:
- Interpreting building information
- Determining quantities
- Relaying information

The following Skills and Knowledge requirements will be covered:
- Understand how to communicate with colleagues and customers in the workplace
- Know company structures, professions and progression routes in the industry
- Know the information and documentation required for completing construction projects
- Known methods of calculating using formulae for construction

The following Certificate learning outcomes will be covered:
- Understand construction information
- Key personnel involved in the day to day communication
- Technical drawings used in the construction industry

The following Diploma learning outcomes will be covered:
- Know how to interpret building information
- Know how to determine building quantities
- Know how to relay information in the workplace

Types of information

This chapter deals with extracting and interpreting information and then correctly relaying it to other people.

Throughout your working life you will have to consult various sources of information and during this chapter you will be required to make decisions and solve problems from the information given or extracted.

Information sources

There are numerous sources of information available and they include the following: drawings, programmes of work, schedules, specifications, policies, mission statements, manufacturers' technical information, organizational documentation, and training and development records and documents.

Drawings

The design team will be required to produce working drawings for the builder to use on the site.

Drawings, schedules and specifications will have to be prepared, explaining how the design team requires the building to be constructed. To be able to read these drawings it is essential that the trainee is able to understand them.

Drawings should be produced according to:

> *BRITISH STANDARD RECOMMENDATIONS FOR DRAWING OFFICE PRACTICE BS 1192*

These recommendations apply to the sizes of drawings, the thickness of lines, dimension of lettering, scales, various projections, graphical symbols, etc.

The person carrying out a task should be able to read drawings and extract the required information.

Information concerning a project is normally given on drawings and written on printed sheets.

Drawings should only contain information that is appropriate to the reader; other information should be produced on schedules, specifications or information sheets.

NOTE

BS 1192 is a British Standard that sets out a best practice method for development, organization and management of production information in architecture, engineering and construction. In 2007 it was updated to BS 1192:2007+A2:2016 to support PAS 1192.

Programmes of work

There are several methods, but the main one used by most construction firms is the bar chart

The bar chart is simple in concept and equally simple to understand. It is hardly surprising therefore that most site supervisors prefer this method.

The activities are listed down the left-hand side in the sequence in which they will take place on site.

The time scale is drawn horizontally and the bars represent the time when the work will proceed on the activities. Bars may be suitably shaded or coloured to distinguish individual trades.

It is usual for only half of the bar to be used for programming; the remainder is completed as progress is achieved. Alternatively, progress can be recorded immediately above or below the bar or in a separate chart specially provided for that purpose.

The person drawing the bar chart could decide the layout, or it could be company policy to adopt a standard format.

The minimum information on the bar chart is the activities and their duration, but plant, labour and material deliveries could also be included.

An example is shown in Figure 3.1. This example shows the activities in their correct sequence. Against each activity is a bar showing the anticipated duration. Each bar should produce a flow through the various activities until completion of the project.

Schedule

This is a contract document that can be used to record repetitive design information about a range of similar components, such as:

- doors
- windows
- ironmongery
- decorative finishes
- inspection chambers (manholes).

A typical schedule for doors is shown in Table 3.1.

Specification

This is a document giving a written description of materials to be used, and construction methods to be employed in the construction of the building (Table 3.2).

No	Activities	Year	2009							
		Month	January				February			
		Week no	1	2	3	4	5	6	7	8
1	Excavate foundation trench		▬							
2	Concrete foundation				▬					
3	Brickwork to DPC					▬				
4	Concrete ground floor slab							▬		
5	External brickwork									▬

FIGURE 3.1
Typical bar chart

Table 3.1 Door schedule

Description	D1	D2	D3	D4	D5
External panelled	•				
Internal flush		•2			
Internal panelled			•		
Internal half-glazed					
Internal glazed					•

Table 3.2 Specification

Item no.	Description
1	Block partition walls, 102.5 in stretcher bond in cement mortar (1:4)
2	Brick facework, 102.5, in stretcher bond using multicoloured facings in cement mortar (1:4) Pointing with half round joint as the work proceeds
3	Blockwork to inner skin of cavity wall, 102.5, in stretcher bond in cement mortar (1.5)
4	Form cavity in hollow wall Insert 50 mm polystyrene batts
5	Damp-proof course, 102.5, of single-layer hessian base bitumen felt and bedded in cement mortar

Policy

A construction company, whether it is a one-person firm or a large company, must have an aim to achieve. To achieve this aim objectives are produced for each person and written down. This group of objectives becomes the policy of the firm.

There will be a different set of policies for the site manager, who has to implement the company's general policy. Typical site policies could be for safety on site, security, quality control, etc.

Mission statement

A mission statement is a formal description of the mission of the business. A typical mission statement is shown in Figure 3.2.

The mission statement might be published on promotional material and on display in the reception area of the company.

It should:

- be brief and easy to understand and remember
- be able to accommodate change
- make the company stand out.

> *Our mission is to be the customer's first choice for affordable housing, delivering outstanding quality and great customer service at a competitive price.*

FIGURE 3.2
Mission statement

Manufacturer's technical information

Any efficient office should have up-to-date information regarding new and existing products from the various manufacturers who produce materials and equipment of interest to them.

Many manufacturers produce technical information which is free for the asking.

The manufacturer's technical information consists of drawings and text which are provided to give the user details of the product. A typical manufacturer's leaflet shown in Figure 3.3 shows technical details for a small mixer.

Organizational documentation

Each type of company will create various organizational documents to be used, but they will usually fall into the following:

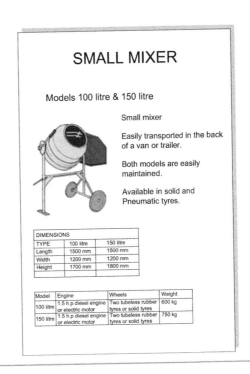

FIGURE 3.3
Manufacturer's leaflet

- time sheets

- job sheet/daywork sheets

- material sheets

- site diary, etc.

Time sheets

Time sheets should be completed weekly by each employee. Failure to complete and hand in a time sheet could result in a loss of wages. A typical time sheet is shown in Figure 3.4.

Daywork sheets

A job sheet or daywork sheet could also be handed out to each employee to record any work completed that was not originally planned.

The RIBA form of contract states:

> 'where work cannot be properly measured and valued, the Contractor shall be allowed daywork rates on the prices prevailing when such work is carried out'.

Records must be kept on a weekly basis.

The site manager must record the number of hours worked, the materials used, and the use of any plant or plant hire, for the duration of the job. A typical daywork sheet is shown in Figure 3.5.

Recording incoming materials

There will be times when you will have to take responsibility to off-load materials, checking that they are exactly what was ordered and of appropriate quality, and then store them properly.

WEST BUILDERS

Timesheet

Employee .... Site ....

Date	Start Time	Finish Time	Total Time	Travel Time	Other expenses
Mon					
Tue					
Wed					
Thu					
Fri					
Sat					
Sun					
Total					

FIGURE 3.4
Time sheet

Daywork Sheet		Contract											
No		Week ending.......................................											
Description of work													
....													
....													

Labour

Name	Trade	Hours						Total hours	Rate	£	P	£	p
		M	T	W	T	F	S						

Materials

Description	Rate	£	p

FIGURE 3.5
Daywork sheet

Always remember to check the delivery ticket before you start to unload the materials. A lorry driver could deliver to the wrong site.

Ensure they are exactly what were ordered by you or the company. A typical delivery sheet is shown in Figure 3.6.

After the driver has off-loaded the materials you will be asked to sign an acceptance slip stating that the order is complete and undamaged.

With certain materials, breakages can be reported to the supplier within a time limit (often 7–14 days) and they will be replaced, but shortages will not. It is therefore very important that shortages are checked before signing.

WEST BUILDERS

Contract :   ....
Contract No : ....
Prepared by: ....

Date	Delivery Note No	Supplier	Materials	Rate	Total value

FIGURE 3.6
Material requisition sheet

If any of the materials or components are missing or defective, they should immediately be identified and noted on both copies of the delivery ticket.

Both tickets should be signed stating the items short or damaged.

If you have found any problems with any delivery you should contact the site supervisor immediately and he will (or will ask you to) contact the supplier to explain the problem.

Site diary

The site supervisor should complete a site diary with various information important to the site. An example is shown in Figure 3.7.

This is one of the most important reports that a site supervisor has to produce during a contract.

It should include the following:

- the weather, morning and afternoon
- any visitors on site
- delays by subcontractors

SITE DIARY AND DAILY REPORT		
Contract	Contract No:	Week ending: ....
Weather	Temperature AM	Temperature PM ....

Labour on site		Plant on site	
Own staff	Subcontractors	Own plant	Hired plant

Job stoppages	Man hours lost	Reason	

Drawings received	Information received	Telephone calls

Visitors to site	Accidents	Reported accidents

Brief report of work and progress on site

Signed :
Site manager ....   Contracts manager ....

FIGURE 3.7
Site diary

- instructions from the clerk of works
- materials delivered
- number of staff on the site.

Training and development records

Any company should consider each employee's career development.

Whatever training has been given it should be documented according to company policy. Probably the first document a new trainee would come into contact with would be the contract of employment, which sets out their terms of employment. A typical contract is shown in Figure 3.8.

Storing information

When information has been found the last place to store it is in your head.

The relative information should be stored in a filing system which everyone can understand. The filing system could consist of something as simple as document files or a more elaborate computer program.

Any information found may be required over and over again for a particular contract, but can also be useful information for future programmes.

The storage of drawings on site can cause problems, especially with dampness. They could be laid flat in a drawing chest or rolled up and placed in racks.

Whichever method chosen they should be recorded when received on site. A typical drawings register is shown in Figure 3.9.

WEST BUILDERS
CONTRACT OF EMPLOYMENT

WB

Statement of main terms of employment

Name of Employer   ....
Name of Employee   ....
Title of Job   ....
Statement issue date   ....
Employment commencement date ....

Your hours of work, rates of pay, overtime, holiday entitlement and payment, pension scheme, disciplinary procedures, notice and termination of employment and disputes procedure are in accordance with the following documents:

1. The National Working Rules
2. The Company Wages Register
3. The Company Handbook

FIGURE 3.8
Contract of employment

WEST BUILDERS Drawings Register					
Contract : .... Drawings from : .... Contract No : .... Date commenced :					
Drawing No	Description	Scale	Number received	Date of issue	Amendments

FIGURE 3.9
Drawings register

Interpreting information

Selecting information from simple drawings, specifications and schedules

It is important that you can understand drawings and extract the correct information from them. The following section gives you a chance to practise reading and understanding drawings, specifications and schedules.

Drawings

You have received the site plan and location plan for a new office block, shown in Figure 3.10. Explain the items marked 1–7 on the plan by completing Table 3.3.

Specifications

These documents are prepared to specify the exact quality of materials and skill required throughout the contract.

It is essential to read and understand exactly what is required on a contract as this will affect the price of the work.

Specifications are used only on large contracts, where the drawings cannot contain all the information required by the contractor.

Drawings for small building works usually have the specification written on the drawings.

A typical specification is shown in Table 3.4. Read it carefully and complete the following questions.

Specification details

To show that you can understand specifications in Table 3.4, complete Table 3.5.

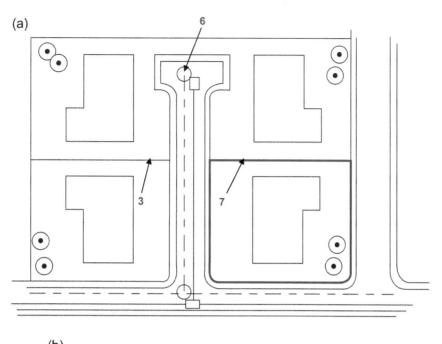

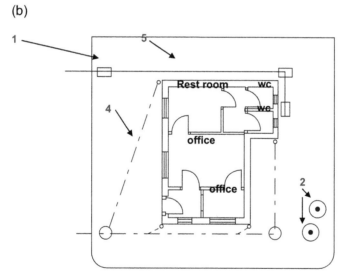

FIGURE 3.10

(a) Site plan and (b) location plan for small office

No.	Definition
1	
2	
3	
4	
5	
6	
7	

Table 3.3 Definitions of items in the plan

Table 3.4 Specification

Item no.	Description
1	Block partition walls, 102.5, in stretcher bond in cement mortar (1:4)
2	Brick facework, 102.5, in stretcher bond using multicoloured facings in cement mortar (1:4) Pointing with half round joint as the work proceeds
3	Blockwork to inner skin of cavity wall, 102.5, in stretcher bond in cement mortar (1:5)
4	Form cavity in hollow wall and insert 50 mm polystyrene batts
5	Damp-proof course, 102.5, of single layer hessian base bitumen felt and bedded in cement mortar
6	Concrete floor slab 150 mm thick, 1:2:4 mix on 2000 gauge DPM on 150 mm consolidated hardcore
7	Screed to concrete floor, 50 mm thick with wooden float finish to receive thermoplastic tiles
8	Internal walls finished with three coat plasterwork
9	Timber flat roof using 200 × 50 joists on 100 × 50 wall plate
10	Window frames fitted according to window schedule as the work proceeds

Table 3.5 Specification details

Question no.	Question	Answer
1	What type of cavity insulation will you order?	
2	What type of damp-proof course is specified?	
3	How many m^2 of blockwork is required?	
4	If three wall ties are required per m^2 how many will you order for the cavity?	
5	Explain the type of mix required for the concrete floor	1: 2: 4:
6	What does DPM mean?	
7	What materials will you order for the screed to the floors?	
8	Will the screed finish be smooth or rough?	
9	Will jigs be required when fixing the windows?	
10	What will be required by the bricklayer when the wall plate is fixed?	

Producing simple working drawings

Scale

Scales are used in drawings to enable large objects such as buildings to be drawn to a convenient size, which will fit onto the drawing paper, while still maintaining accurate proportions which can be drawn or measured as required.

Scales use ratios to relate measurements on a drawing to the real dimensions of the actual item being drawn.

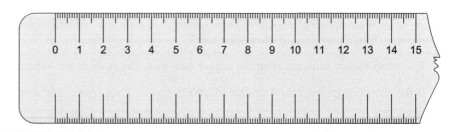

FIGURE 3.11
Ruler

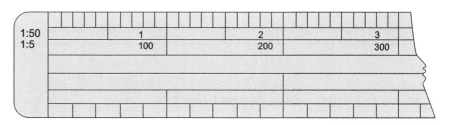

FIGURE 3.12
Scale rule

Once these items have been drawn to a smaller size, which is known as a ratio to the real item, it is known as a scale drawing.

Scale rules

An ordinary ruler, 300 mm long (Figure 3.11), is useful for simple drawings, but scale drawings require a scale rule.

A scale rule (Figure 3.12) has a series of marks for measuring purposes. The scale rule is graduated in millimetres. Each scale represents a ratio of given units to one unit.

Table 3.6 lists the preferred scales for building drawings.

Use of scales

The choice of scale will depend on two things:

- the size of the object to be drawn
- the amount of detail that needs to be shown.

Table 3.6 Preferred scales for building drawings	
Type of drawing	Scales
Block plans	1:2500, 1:1250
Site plans	1:500, 1:200
Location drawings	1:200, 1:100, 1:50
Range drawings	1:100, 1:50, 1:20
Detail drawings	1:10, 1:5, 1:1
Assembly drawings	1:20, 1:10, 1:5

A scale is used to measure distances on drawings and for taking measurements of a drawing. If scaled dimensions and written dimensions disagree, then the written dimension should always be used.

The ratio shows how many times bigger one quantity is than the other.

If the drawing is made to a scale of 1:100, 1 mm will represent 100 mm.

If the drawing is made to a scale of 1:20, 1 mm will represent 20 mm, and so on.

If the following drawing was drawn 150 mm long it would represent something much bigger in size:

To a scale of 1:10 it would represent 1500 mm.

If it is drawn to a scale of 1:20 it would represent 3000 mm.

All you have to do is multiply the scale measurement by the scale ratio. Scale rules can be used for both reading and preparing a drawing.

Producing simple drawings

There are many methods of drawing a particular object, each of which can portray a different feature of the object.

Before a drawing is started you must decide what you actually need to define.

You need to decide whether a front, side or back view is required, or even a plan of the object.

Differing perspectives are also possible, along with sections and exploded views.

A decision must also be made on whether the finished drawing will be a traditional line drawing or a freehand drawing.

Basic drawing methods

Methods of projection

There are several methods of projection, the most common of which are:

- orthographic projection
- isometric projection
- oblique projection.

Orthographic projection is the method of showing solid objects which are actually three-dimensional in two-dimensional drawings by means of related views and plans, elevations and sections.

> **Remember**
>
> It is recommended that written dimensions on a drawing should always be used, rather than taking dimensions off drawings with a scale rule.

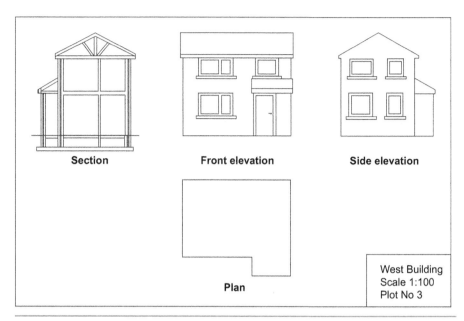

FIGURE 3.13
Orthographic projection

Most constructional drawings are of this nature. A typical example is shown in Figure 3.13.

The projected faces that represent the sides of the object are known as elevations. Those viewed looking down from above or up from below are known as plans.

The outline of the building shown can be viewed in the direction of the arrows to show all the relevant details.

The views should be set out on drawing paper in a special format.

The front elevation is drawn first and the other views are related to it.

The plan of the building is always drawn directly underneath the front elevation.

The view from the left is placed to the right of the front elevation and the view from the right is placed to the left of the front elevation.

A view from the rear may be placed to the left or right depending on space and personal choice.

A section is placed on the opposite side to the intended view.

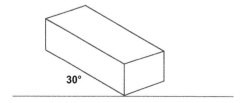

FIGURE 3.14
Isometric projection

Isometric projection

This is a method of presenting the detail in pictorial format (Figure 3.14).

The definition of isometric projection is that all vertical lines remain the same but the horizontal lines are all drawn at 30 degrees to the horizontal.

Oblique projection

This is another method of presenting the detail in pictorial format.

There are two main types of oblique projections, cavalier and cabinet (Figure 3.15). Both are drawn with the front face drawn to true shape and size, similar to a front elevation.

The faces that recede are drawn at an angle of 45 degrees, and the difference between the two methods is the scale of these receding lines. The cavalier method uses the full scale, whereas the cabinet uses half full size scale, which creates a more normal appearance.

Scaled drawings

The preceding pages have dealt with the various techniques involved in construction drawings. This section examines in more detail the types of construction drawings that you will have to prepare in your assignments.

All the points raised before will also apply to the drawings.

Restrictions have been placed on the figures by the page size; the actual sizes of production drawings are likely to be much larger.

The following drawings would be found within most construction drawings:

- elevations
- plans
- sections.

Most construction drawings are scale drawings, that is, the land, buildings or objects shown on them are not represented at their true size, but are larger or more often smaller in some proportion.

Drawings for a small garage are shown in Figure 3.16.

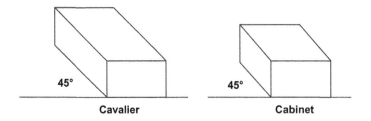

FIGURE 3.15
Oblique projections: (a) cavalier; (b) cabinet

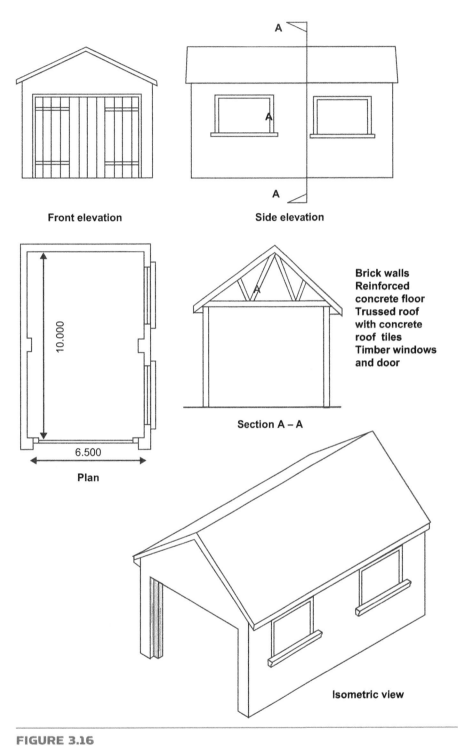

Front elevation

Side elevation

10.000

6.500

Plan

Brick walls
Reinforced
concrete floor
Trussed roof
with concrete
roof tiles
Timber windows
and door

Section A – A

Isometric view

FIGURE 3.16
Typical drawing layout

The position of drawings on the paper depends on the size of the contract, but in general the plans should always be underneath the elevations, with the sections at the side of the elevations.

On small contracts all the drawings could be positioned on one sheet of paper, but on larger contracts, where several hundreds of details are required,

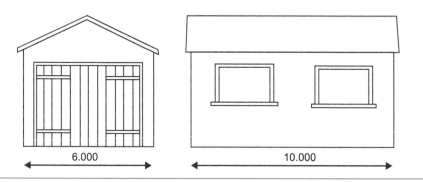

FIGURE 3.17
Elevations of the garage

numerous drawing sheets are required. They should all be correctly numbered as received.

Elevations

These consist of front elevation, side elevation and occasionally rear elevation. Elevations of the garage are shown in Figure 3.17.

They are drawn to a scale of 1:200, 1:100 or 1:50, depending on the detail required. The elevations give a visual effect of the finished building from all directions.

They have limited dimensions on them as most information is gained from other drawings.

Plans

These are drawn to a scale of 1:200, 1:100 or 1:50, usually to match the elevations. The plan of the garage is shown in Figure 3.18.

As stated before, the ground floor plan should be drawn below the front elevation. If there is room, the first floor plan should be drawn below the ground floor plan.

One of the main purposes of the floor plans is to show the disposition of the various rooms, the position of entrances, circulation routes and the overall character of the building.

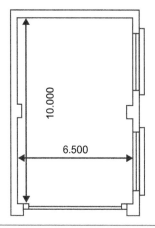

FIGURE 3.18
Plan of the garage

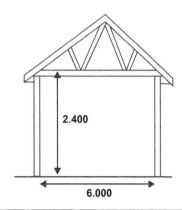

FIGURE 3.19
Section

The size is still too small to add too much in the way of dimensions.

Notes, if included on such drawings, tend to be general descriptions.

Sections

Sections of the building are chosen to display a detail that otherwise would be difficult to see on the plans and elevations.

These drawings can either be cross-sectional or longitudinal sections, and either full or part sections. A section of the garage is shown in Figure 3.19.

A full cross-section would show the full height of the building and its respective floor and ceiling levels.

Depending on the scale chosen, information would be restricted to that which could not be found elsewhere.

Quantities of materials

You will at some stage in your career have to determine quantities of materials required to carry out an activity. This will entail having to calculate areas and volumes.

The buyer will have the responsibility of taking off and scheduling materials from the bills of quantities, or on smaller jobs, the working drawings.

When the site supervisor is given this role he or she must be careful to ensure that correct quantities and details of quality of materials are extracted from the drawings and specifications.

Allowances should be made for materials wastage which, from experience, is usually 5–10 per cent.

A bricklayer should know how to work out the number of bricks needed to build a wall. This is a very simple matter using a pocket calculator. Do not panic at the very mention of arithmetic. Let the calculator take the strain and deal with the decimal point!

Method A

There is nothing wrong with counting up the number of bricks needed for one course, then multiplying by the total number of courses in the wall to get the answer.

This method works well for stretcher bond walling, but becomes difficult when applied to more complicated shapes that have doors and windows, or to cavity walling.

Method B

This method is based on finding the surface area of the wall. It makes deductions for doors and windows easy. Use the calculator to add, subtract and multiply. Do not strain the brain!

Types of calculation

Calculations in construction can be:

- numbers – one-off items such as chimney pots
- linear – lengths of materials, etc.
- superficial – areas of walls and floors, etc.
- cubic – volume of excavations, etc.
- time – used for calculating bills
- cost – used in pricing materials and labour.

Number

Many building materials are measured by number.

Chimney pots, air bricks, doors and windows are all examples of materials that are measured by number.

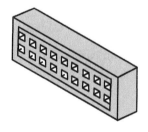

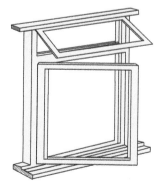

Linear

Many building materials are measured by length, but stating the width and thickness.

Timber is measured by length, for Example 3 m length of 50 × 50 mm.

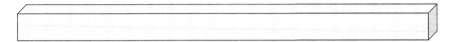

Superficial

This is also known as square measurement, as the length and depth are multiplied to achieve the superficial area of the material required.

Examples are brick and blockwork.

The length of the wall is multiplied by the height to give the square area of brickwork required.

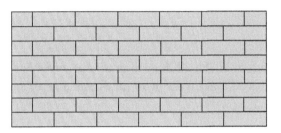

Cubic

Cubic measurements are taken when there are three dimensions: length, width and depth.

Examples are found in excavating trenches and concrete for foundations.

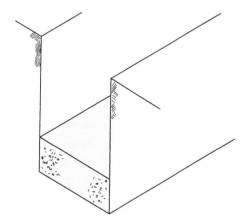

Time

Time is used by all construction firms when calculating bills using information from time sheets completed by each employee.

Cost

Estimating the cost of the work is impossible without knowledge of money. All employees in the construction industry are employed on a certain rate for the job. Calculating the cost for any work will involve both time and cost.

Calculators

These are a very useful tool, but do not rely on the calculator at first.

A basic calculator with +, -, ×, ÷, e is all that is needed.

Memory in a calculator can also be useful, but all steps need to be shown on paper.

It is particularly important to check answers when using calculators because it is easy to press the wrong button on small calculators.

Decimals

It is important that the decimal points are correctly placed. Misplacing a decimal point can make a result 10 times too large or small.

The metric system

The metric system has been devised so that standard units are used throughout the world. The metric system is much easier to use in calculations and measurements than the imperial system.

Unit of length

metre (m) millimetre (mm) kilometre (km)

1 km = 1000 m

1 m = 1000 mm

Therefore 1 km = 1 000 000 mm.

Unit of mass

kilogram (kg)

gram (g)

tonne (t)

1 t = 1000 kg

1 kg = 1000 g

Unit of force

Newton (N)

> **Remember**
>
> All calculations should be shown and not done on bits of scrap paper or in your head. This will enable any mistakes to be found.
>
> When end tests are being carried out marks are given for the correct method of setting out and not just for the answer.

kilonewton (kN)

meganewton (MN)

1 MN = 1000 kN

1 kN = 1000 N

To convert mass (weight) to force = multiply by 9.81 m/s.

Examples

Bricks

The drawing below shows the nominal sizes of a brick as:

215 mm × 102:5 mm × 65 mm

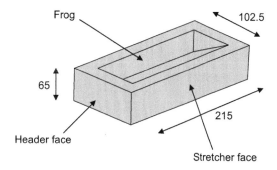

These are the net sizes of the average brick and it is necessary to add 10 mm to the length, width and depth to produce a size when installed in the wall.

Therefore: length = 225 mm; depth = 75 mm and width

= 112:5 mm:

Bricks are laid to a gauge (number of courses to a set measurement) of four courses to 300 mm.

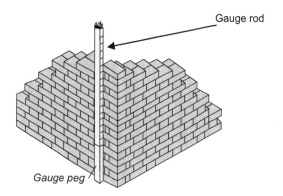

Brick calculations

The size of a brick including the joints:

= 225 × 112:5 × 75 mm

If we find the face area of one brick and divide that into 1 square metre it will result in the brick requirement.

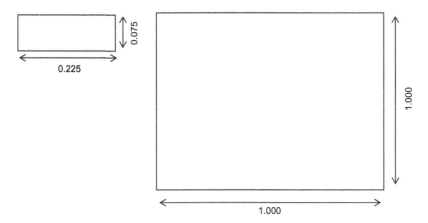

Area of the stretcher face of a brick:

= 0:225 × 0:075 = 0:016875 m^2

Area of a half-brick wall = 1 m × 1 m = 1 m^2

Therefore,

0:01678 m^2 = 59:27 bricks

Allowing for waste = 60 bricks per square metre of half-brick thick wall. The other brick requirements are:

- 1 brick wall: 120 bricks

- 1^1/$_2$ brick wall: 180 bricks.

Mortar calculations

Nominal size for a brick is 215 × 102.5 × 65 mm.

Nominal size for a joint is 10 mm.

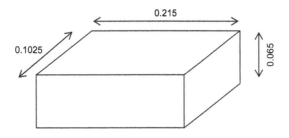

If the brick size = 215 × 102.5 × 65 mm, then

Bed joint = Length × Width × Thickness of bed joint

= 0.125 × 0.1025 × 0.01

= 0.00013 m^3

Cross-joint = Width × Depth × Thickness of cross-joint

= 0.1025 × 0.075 × 0.01

= 0.00008 m^3

Therefore, the total for one bed joint and one cross-joint

= 0.00013 + 0.00008

= 0.0002 m^3

If the calculation was for 1000 bricks,

Mortar required = 0.0002 × 1000

= 0.2 m^3

Therefore,

$^1/_2$ brick wall = 60 bricks + 60 kg mortar

1 brick wall = 120 bricks + 120 kg mortar

$1^1/_2$ brick wall = 180 bricks + 180 kg mortar

Example 1

Calculate the number of bricks required for the wall area.

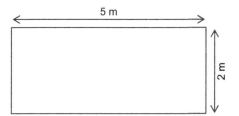

5 m

2 m

Note

Normal allowance is around 0.6 m^3 per 1000 bricks, which allows for the frogs and waste.

Another method to simplify the calculation is to allow 1 kg of mortar for each brick.

It follows that 1000 bricks would require 1000 kg of mortar, which is 1 tonne.

Answer 1

Surface area = 5 × 2 = 10 m^2 (square metres)

You require 60 bricks (from previous calculations) to build 1 m^2 of $^1/_2$ brick thick (102 mm) stretcher bond walling.

The surface area of the wall is 10 m^2, so you need 10 × 60 = 600 bricks.

Example 2

Calculate the number of bricks required for the wall area.

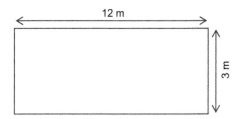

12 m

3 m

Answer 2

Surface area = 12 × 3 = 36 m^2 (square metres)

You require 60 bricks to build 1 m^2 of $^1/_2$ brick thick (102 mm) stretcher bond walling.

The surface area of the wall is 36 m^2, so you need 36 × 60 = 2160 bricks.

Walls one brick (1B) thick (215 mm) need 120 bricks to build 1 m^2 (1 m^2), and the same working method can be used as shown in Example 1.

Example 3

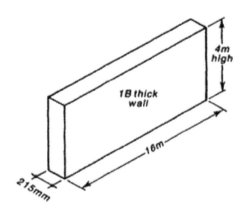

How many bricks are needed for this wall?

Answer 3

120 bricks per m^2

Two lines of working method:

Surface area = 16 × 4 = 64 m^2

Number of bricks needed = 64 × 120 = 7680 bricks

Walls are not always straight elevations, and buildings have corners, but the same method of calculation can be used.

Example 4

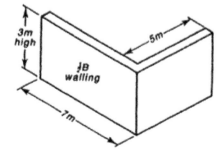

How many bricks are required to build the wall shown?

Answer 4

60 bricks per m^2

Three lines of working method:

Total length of wall = 7 + 5 = 12 m

Surface area = 12 × 3 = 36 m^2

Number of bricks required = 36 × 60 = 2160 bricks

Note

If calculations are not a happy thought for you, it is always a good idea to retrace your steps. Go back to Example 1.

Just change the dimensions of that wall, invent your own figures, and then work out how many bricks will be needed this time.

Repeat what you have just done, but with Examples 2, 3 and 4, to calculate the number of bricks required for each example, but with different wall dimensions.

Example 5

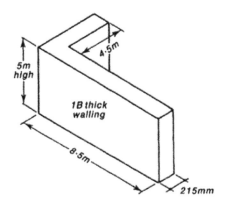

How many bricks are required to build this wall?

Answer 5

120 bricks per m^2

Three lines of working method:

Total length = 8.5 + 4.5 = 13 m

Surface area = 13 × 5 = 65 m^2

Number of bricks required = 65 × 120 = 7800 bricks

Example 6

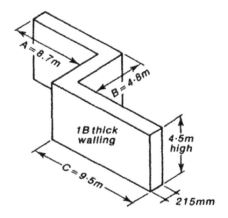

How many bricks are required to build this wall?

Answer 6

120 bricks per m^2

Three lines of working method

Total length = A + B + C = 8.7 + 4.8 + 9.5 = 23 m

Surface area = 23 × 4.5 = 103.5 m^2

Number of bricks required = 103.5 × 120 = 12420 bricks

Blocks

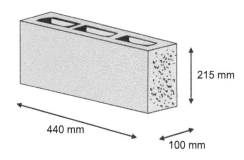

Note

For additional practice, repea
Examples 5 and 6 after changin
the dimensions, invent your own
and assume that the walls are onl
102 mm thick.

The dimensions for blocks are shown as:

440 mm long × 215 mm high × 100 mm wide

These are net dimensions and the thickness of the joint is added to obtain the following dimensions:

450 mm long × 225 mm high × 100 mm wide

Blocks when laid are equal to three courses of bricks.

The calculations for blocks are very similar to brick calculations.

First calculate the area of one block then divide it into 1 square metre.

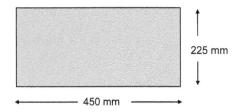

450 mm × 225 mm = 0.10125 mm^2

Number of block per square metre=1.000/0.10125 = 9.87 blocks

It is usual to include waste and allow 10 blocks per square metre.

Blockwork

Example 7

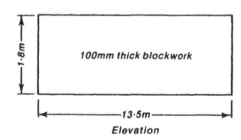

Calculate how many standard size blocks are required to build this wall.

Answer 7

10 blocks per m^2 (from previous calculations) Two lines of working method:

Surface area = 13.5 × 1.8 = 24.3 m^2

Number of blocks = 24.3 × 10 = 243 blocks

Example 8

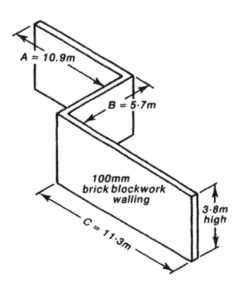

How many standard size blocks are needed to build this wall?

Answer 8

10 blocks per m^2

Three lines of working method: Total length of wall = A + B + C

= 10.9 + 5.7 + 11.3 = 27.9 m

Surface area = 27.9 × 3.8 = 106.02 m^2

Number of blocks = 106.02 × 10 = 1060.2

Rounded off to 1061 blocks

Mortar

Allow 0.02 m³ (cubic metre) (from previous calculations) of ready-mixed mortar to build 1 m² (square metre) of stretcher bonded brickwork. This is an average figure, bearing in mind that some bricks have frogs and others are perforated.

Table 3.7 shows approximately how much ready-mixed mortar is required for different wall thicknesses in brick and block.

Note

When ordering blocks for the inne leaf of cavity walls or for partition walls, take care to make clear the thickness you want, 100, 140, 150 190 or 215 mm, as well as type compressive strength, density and surface finish.

Example 9

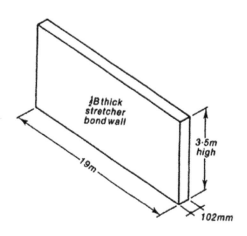

Calculate how much mortar will be required to build the brick wall.

Answer 9

0.02 m³ of mortar per m² of walling (from previous calculations)

Two lines of working method:

Surface area of wall = 19 × 3.5 = 66.5 m²

Volume of mortar required = 66.5 × 0.02 = 1.33 m³

Note

For additional practice, go back to calculate how much mortar would be required to build the wall in Example 4.

Example 10

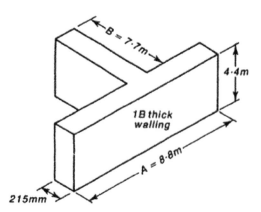

Calculate how many bricks and how much mortar will be required to build the following wall:

120 bricks and 0.05 m³ of mortar per m²

Table 3.7 Materials per square metre of walling

	Stretcher bond			English bond			Flemish bond			English garden wall bond			Flemish garden wall bond			Header bond		
	F	C	M (m³)	F	C	M (m³)	F	C	M (m³m³)	F	C	M (m³)	F	C	M (m³)	F	C	M (m³)
½B 102 mm thick	60		0.02															
1B 215 mm thick	60ᵃ	60ᵃ	0.05	90	30	0.05	80	40	0.05	73 ᵇ	47	0.05	67 ᵇ	53	0.05	120		0.047
1½B 327 mm thick				90	90	0.08	80	100	0.08	73	107	0.08	67	113	0.08			
2B 440 mm thick				90	150	0.11	80	160	0.11	73	167	0.11	67	173	0.11			
100 mm standard size blocks	10		0.01															
140 mm standard size blocks	10		0.014															
150 mm standard size blocks	10		0.015															
190 mm standard size blocks	10		0.019															
215 mm standard size blocks	10		0.022															

F: facings; C: commons; M: mortar.

Notes

a 215 mm thick walls can be built to show stretcher bond and give a face finish if required both sides if steel mesh bed joint reinforcement or butterfly wall ties are incorporated every fourth course.

b Garden wall bonds are intended for 215 mm thick free-standing boundary walls using 100% facing bricks.

Answer 10

Four lines of working method:

Total length = A + B = 8.8 + 7.7 = 16.5 m

Surface area = 16.5 × 4.4 = 72.6 m^2

Number of bricks = 72.6 × 120 = 8712 bricks

Volume of mortar = 72.6 × 0.05 = 3.63 m^3

Example 11

Calculate how many standard sized blocks and how much mortar will be required to build the following block wall:

10 blocks and 0.01 m^3 of mortar per m^2

Answer 11

Four lines of working method:

Total length = A + B + C + D

= 11.4 + 8.2 + 9.6 + 7.3 = 36.5 m

Surface area = 36.5 × 2.7 = 98.55 m^2

Number of blocks = 98.55 × 10 = 985.5 blocks

Rounded off to 986 blocks

Volume of mortar = 98.55 × 0.01 = 0.9855 m^3

Rounded off to 1 m^3

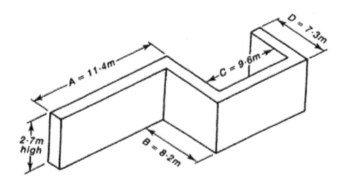

Rounding off

With building calculations, it is customary to 'round off' decimal points, so as to tidy up quantities of materials into whole numbers where possible.

For example, it is not practical to order half a bag of cement, nor worry about 0.2 of a cubic metre of a bulk item like trench excavation.

It is sensible to round up with building materials, as site handling can result in damage or loss even in excess of the percentage allowances for cutting and waste.

Taking Examples 11, 1 m^3 (1 cubic metre) is a less cumbersome figure than 0.9855 m^3 and involves fewer numbers with which to make mistakes.

Openings

Most walls and buildings have door and window openings. This means that you must reduce your order for bricks and blocks, depending on the size and the number of openings.

The best way is to deduct the surface area of an opening from the overall wall area, before the final step of calculating the number of bricks or blocks needed (see Example 12).

Example 12

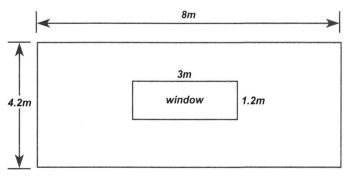

Elevation of $\frac{1}{2}$B wall

Calculate how many bricks will be required to build this wall, making the necessary deduction for the opening.

Answer 12

Using 60 bricks per m^2

Four lines of working method:

Overall surface area of the wall = 8 × 4.2 = 33.6 m^2

Area of window opening = 3 × 1.2 = 3.6 m^2

Net area of walling = 33.6 - 3.6 = 30 m^2

Number of bricks required = 30 × 60 = 1800 bricks

Example 13

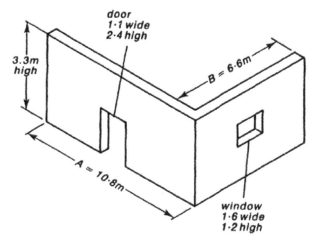

How many bricks are required to build the wall shown, making allowances for openings?

Answer 13

Using 60 bricks per m^2

Five lines of working method:

Total length of wall = A + B

= 10.8 + 6.6. = 17.4 m

Surface area = 17.4 × 3.3 = 57.42 m^2

Area of door opening = 1.1 × 2.4 = 2.64 m^2

Area of window opening = 1.6 × 1.2 = 1.92 m^2

Net area of walling = 57.42 - 4.56 = 52.86 m^2

Number of bricks required = 52.86 × 60 = 3171.6 bricks

Rounded off to 3172 bricks

Note

For additional practice, calculate how much ready-mixed mortar will be required to build each wall shown in Examples 12 and 13. (Follow lines of working method given in Example 9.)

Cavity walling

When working out quantities of materials for cavity work, each leaf of the walling must be dealt with on its own, because you need separate totals for ordering bricks and blocks (see Example 14).

Example 14

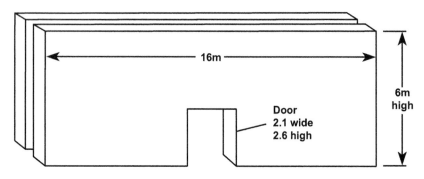

Elevation of 265mm thick cavity wall

Calculate how many facing bricks and blocks will be required to build the cavity shown, making due allowances for the opening.

Answer 14

Using 60 bricks and 10 blocks per m^2

Five lines of working method:

Overall surface area = 16 × 6 = 96 m^2

Area of opening = 2.1 × 2.6 = 5.46 m²

Net area of walling = 96 - 5.46 = 90.54 m²

Number of facing bricks = 90.54 × 60 = 5432.4

Rounded off to 5433 bricks

Number of blocks = 90.54 × 10 = 905.4

Rounded off to 906 blocks

Perimeters

Perimeter of a rectangle = (2 × Length) + (2 × Width)

Example 15

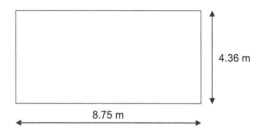

Find the perimeter of the area shown.

Answer 15

Perimeter = 2 × L + 2 × W

= (2 × 8.75) + (2 × 4.36)

= 17.50 + 8.72

= 26.22 m

Example 16

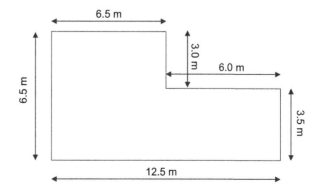

Find the perimeter of the area shown.

Answer 16

Perimeter = 2 L × 2 W

= (2 × 12.5) + (2 × 6.5)

= 25 + 13

= 38 m

There is no need to add all the individual sides when the overall length and width is known. If the dimension was missing from the long side then the two smaller sides would have to be added together to give the overall length

= 6:5 + 6 = 12:5 m

Example 17

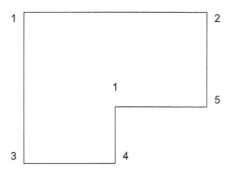

The first design shows five external corners and one internal corner. If you deduct the internal from external corners you are left with four external corners.

This rule applies to all rectangular designs.

Example 18

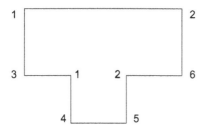

The second design shows six external and two internal corners. If you deduct the internal from external corners you are still left with four external corners.

This rule applies to all rectangular designs with corner insets.

Remember also that the perimeters of designs with the same overall dimensions are the same irrespective of the number of corner insets.

Example 19

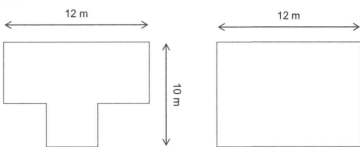

The two designs have the same perimeters even if the shapes are different.

The only time the perimeters are different and the number of external angles is more than four is when the design incorporates insets.

Example 20

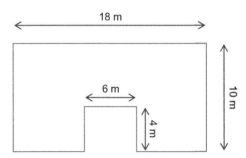

Answer 20

The perimeter is equal to that of a rectangle with sides of 18 and 10 plus the depth of each insert line.

Therefore:

Perimeter = (2 × 18) + (2 × 10) + (2 × 4) = 64 m

Method C: The centre line method

When walls have many more returns than Example 11, it can be difficult to ensure that corners are not measured twice over. Professional quantity surveyors make use of the 'centre line method' to overcome this problem when measuring up walls.

This simple but effective idea means that you first work out the total length on plan, of the centre line, of all the intersecting walls of a building having the same thickness. This figure is then used to calculate the total area of walling in one single sum (see Example 21).

The previous two examples explained how the external perimeter of a building is calculated, but it is necessary to move this line into the centre of each wall to ensure that the calculation is correct.

This is done by making a deduction for each external angle equivalent to the thickness of each wall.

If the dimensions were internal then an addition could be made for each internal angle equal to the thickness of each wall.

Example 21

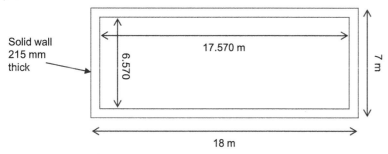

Solid wall
215 mm
thick

17.570 m

6.570

7 m

18 m

Calculate the centre line for the building with a solid brick wall.

Answer 21

Taking external dimensions:

Total external dimensions = (18 + 7) × 2 = 50 m

Deduct corners = 4 × 215 = 0.860 m

Centre line of solid brick wall = 50 - 0.860 = 49.140 m

Taking internal dimensions:

Total internal dimensions = (17.570 + 6.570) × 2 = 48.280 m

Add corners = 4 × 215 = 0.860 m

Centre line of solid brick wall = 48.280 + 0.860 = 49.140 m

The following enlarged drawing will explain in more detail why it is necessary either to add or to deduct the full thickness when adjusting for corners.

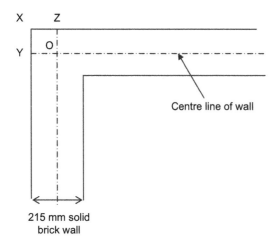

X   Z

Y

O

Centre line of wall

215 mm solid
brick wall

The drawing shows one corner of a 215 mm solid brick wall. The centre line passes through the intersection point at 'O'.

The centre lines both extend to meet the outside face of the wall at Y and Z.

This shows quite clearly that the lines XY and XZ need to be deducted to produce the correct centre line.

The two dimensions are equivalent to two half walls which together add up to a full wall thickness. There are always four corners which need either deducting when taking external dimensions or adding when taking internal dimensions.

When using the centre line method on solid walls it gives the centre line for the excavation, the foundation concrete, the damp-proof course (DPC) and the brickwork. So you can see that it is a very important calculation.

When cavity wall construction is being measured the calculation is a little more complicated but the same principle applies.

The centre line of the wall will be required for the excavation, foundation concrete and formation of the cavity and possible cavity insulation, but a deduction or addition will need to be made to calculate the centre line of the external facings and the inner blockwork.

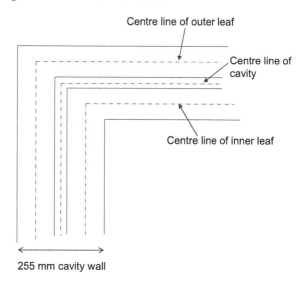

Assume the external wall to be 102.5 mm wide, the cavity 50 mm and the inner blockwork 102.5 mm.

To arrive at the centre line of the face brickwork, taking external dimensions, you would need to deduct 4 × 102.5 from the overall perimeter dimensions.

To calculate the centre line of the cavity you first need to deduct the same figure (4 × 102.5) to arrive at the inside face of the outer leaf of face brickwork. Then deduct 4 × 50 to find the centre line of the cavity.

To find the centre line of the blockwork you first need to move the centre line to the inner face of the blockwork by deducting 4 × 50 mm cavity. Deduct 4 × 102.5 to find the centre line of the blockwork.

Example 22

Calculate the centre line for both the outer leaf of facing bricks and the inner leaf of blocks.

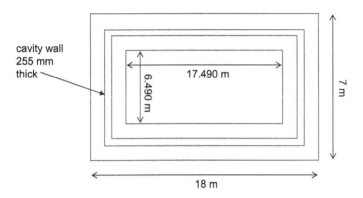

Answer 22 a: External dimensions

Centre line of outer leaf of facing bricks:

Total external dimensions = (18 + 7) × 2 = 50 m

Deduct corners = 4 × 102.5 = 0.410 m

Centre line of facing brick outer leaf = 50 - 0.410 = 49.59 m

Centre line of inner face of facing bricks:

Centre line of outer leaf = 49.59 m

Deduct 4 × 102.5 = 0.410 m

Inner face of brick outer leaf = 49.59 - 0.410 = 49.18 m

Centre of cavity:

Centre line of inner face of facing bricks = 49.18 m

Deduct 4 × 50 = 0.200 m

Centre line of cavity = 49.18 - 0.200 = 48.98 m

Inner face of blockwork:

Centre line of cavity = 48.98 m

Deduct 4 × 50 = 0.200 m

Centre line of inner face of blockwork = 48.89 - 0.200 = 48.78 m

Centre line of blockwork:

Centre line of inner face of blockwork = 48.78 m

Deduct 4 × 102.5 = 0.410 m

Centre line of inner leaf of blockwork = 48.78 - 0.410 = 48.37 m

Answer 22b: Internal dimensions

Centre line of inner leaf of blockwork:

Total internal dimensions = (17.49 + 6.49) × 2 = 47.96 m

Add corners = 4 × 102.5 = 0.410 m

Centre line of blockwork inner leaf = 47.96 + 0.410 = 48.37 m

Centre line of inner face of blockwork:

Centre line of inner leaf = 48.37 m

Add 4 × 102.5 = 0.410 m

Inner face of blockwork inner leaf = 48.37 + 0.410 = 48.78 m

Centre of cavity:

Centre line of inner face of blockwork = 48.78 m

Add 4 × 50 = 0.200 m

Centre line of cavity = 48.78 - 0.200 = 48.98 m

Inner face of brickwork:

Centre line of cavity = 48.98 m

Add 4 × 50 = 0.200 m

Centre line of inner face of brickwork = 48.89 + 0.200 = 49.18 m

Centre line of brickwork:

Centre line of inner face of brickwork = 49.18 m

Add 4 × 102.5 = 0.410 m

Centre line of outer leaf of brickwork = 49.18 + 0.410 = 49.59 m

If you check the above answers they will be the same irrespective which method is used.

If we combine the centre line method and the material calculation we should arrive at the total materials required for any building.

Example 23

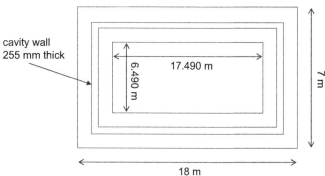

Using the centre lines calculated in Example 22, calculate the facing bricks and blocks required for the building, assuming the walls are 2.2 m high.

Answer 23

Using 60 bricks and 10 blocks per m²

Number of facing bricks:

Centre line of facing bricks × 2.2 = 49.59 × 2.2 = 109.098 m²

Number of facings = 60 × 109.098 = 6545.88 bricks

Rounded off to 6546 bricks Number of blocks:

Centre line of blockwork × 2.2 = 48.37 × 2.2 = 106.414 m²

Number of blocks = 10 × 106.414 = 1064.14 blocks

Rounded off to 1065 blocks

Example 24

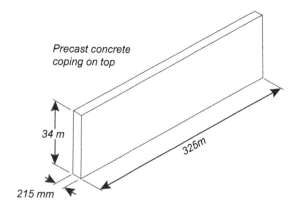

Calculate the total area for repointing the brick wall, back, front and ends.

Answer 24

Perimeter = (32.5 m) × 2 + (0.215 × 2)

= 65 m + 0.43 m

= 65.43 m

Total area of repointing = 65.43 × 3.4 = 222.462 m²

Rounded off to 233 m²

Example 25

Calculate the total area of raking out joints and repointing external walls of the building shown. (Ignore door and window openings for this example.)

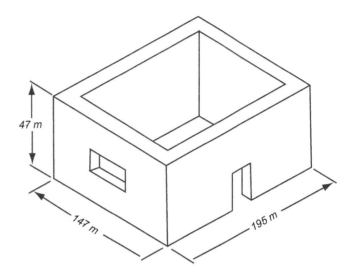

Answer 25

Perimeter = (19.5 m) × 2 + (14.7 × 2) = 39 m + 29.4 m = 68.4 m

Total area of repointing = External perimeter × 4.7 = 68.4 × 4.7 = 321.48 m^2

Rounded off to 322 m^2

Division of facings and commons. The outer, 102 mm thick leaf of cavity work and other half brick walling is usually built in stretcher bond, so that all the bricks required will be facings.

With 215 mm and thicker solid brickwork faced on one side only, some bricks will need to be less expensive commons to back up the facing bricks.

With 215 mm thick Flemish bond walling, 2/3 will be facings and 1/3 commons. With 215 mm English bond walling, 3/4 will be facings and 1/4 will be commons.

If 215 mm thick walls are built to header bond, then of course all bricks ordered should be facings. This information is summarized in Table 3.7.

Note

For extra practice repeat Examples 24 and 25 after changing the dimensions (invent your own).

Example 26

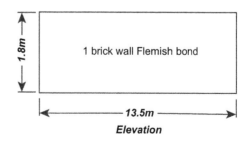

1 brick wall Flemish bond

1.8m

13.5m

Elevation

Assume that the wall shown is to be 215 mm thick Flemish bonded brickwork.

Calculate the number of facings and commons that will be required, in separate totals.

Answer 26

Taking information from Table 3.7:

80 facings per m²

40 commons per m²

Three lines of working method:

Surface area = 13.5 m × 1.8 m = 24.3 m²

Number of facings = 24.3 × 80 = 1944 facing bricks

Number of commons = 24.3 × 40 = 972 common bricks

Percentage for cutting and waste

All of the foregoing examples of calculating requirements for bricks, blocks and mortar have resulted in exact or net quantities. Owing to breakages and waste when cutting and handling bricks and blocks and when using mortar, it is necessary to increase net orders by 5 per cent.

Therefore as an additional stage at the end of each calculation, 5 per cent extra should be added on, using the % button on your calculator, for the purposes of ordering.

Example 27

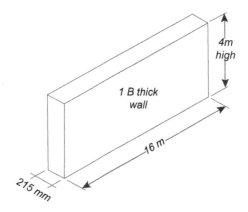

Calculate the amount of facing bricks and common bricks for the wall shown if English bond was used.

Allow 5 per cent for wastage

Answer 27

Taking information from Table 3.7: 90 bricks per m²

30 commons per m²

Two lines of working method:

Surface area = 16 × 4 = 64 m²

Number of facings = 64 × 90 = 5760 facing bricks

Add 5% for wastage = 5760 × 5 O 100 = 288

Therefore, total facing bricks required = 5760 + 288 = 6048 facing bricks

Number of commons = 64 × 30 = 1920 commons

Add 5% for wastage = 1920 × 5 O 100 = 96

Therefore total commons required = 1920 + 96 = 2016 common bricks

Note

Go back to Example 23 and calculate the number of bricks and blocks that should be ordered after allowing 5 per cent for cutting and waste.

Example 28

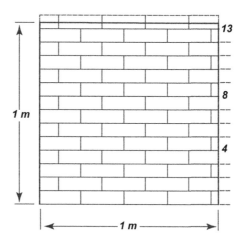

Calculate the area of mortar joints (say to assess the effect of mortar colour on finished walling) showing on the surface of 1 m^2 of stretcher bonded brickwork.

For the purposes of this calculation, assume that the elevation above is 'stack bonded'.

Answer 28

Bed joint mortar showing = 1 m × 0.01 × 13 = 0.13 m^2

Cross joint mortar showing = 1 m × 0.01 × 5 = 0.05 m^2

Total = 0.18 m^2

Expressed as a percentage of 1 m^2 = 0.18/1 m^2 × 100 = 18% of surface viewed is mortar colour. The remaining 82% only is brick colour.

Additional applications of the centre line method

In addition to bricks, blocks and mortar, a bricklayer may be asked to work out quantities of cavity insulation, DPC, foundation concrete or even excavation to trenches.

Each such calculation is very simple, if you find out the total length of the centre line of the walling first.

Example 29

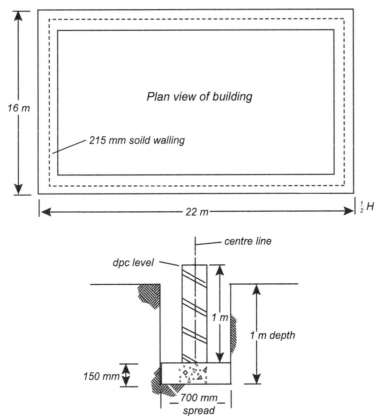

Answer 29

Centre line

Total external dimensions = $(22 + 16) \times 2 = 76$ m

Deduct corners = 4×215 m = 0.860 m

Centre line of solid brick wall = 76 - 0.860 = 75.14 m

The dotted centre line perimeter shown on the drawing is exactly 75.14 m long. This dotted line is the exact centre line of the trench excavation, concrete strip foundation, substructure brickwork and DPC, all with a net length of exactly 75.14 m that can be used in four separate calculations.

Trench excavation

Volume = Centre line × Width × Depth

= 75.14 m × 0.70 m × 1 m = 52.598 m^3

Rounded off: 53 m^3

Strip foundation concrete

Volume = Centre line × Spread × Depth

= 75.14 m × 0.70 m × 0.150 m

= 7.890 m^3

Rounded off: 8 m^3

Substructure brickwork

Area = Centre line × Height

= 75.14 m × 1 m = 75.14 m^2

Number of bricks = 75.14 × 120 = 901.68

Rounded off: 902 bricks

DPC

Linear measurement = Centre line + Extra for laps

= 75.14 + 5% extra for laps = 75.14 × 5 O 100 = 3.757

Therefore total DPC required = 75.14 + 3.757 = 78.897

Rounded off to 79 m length of 215 mm wide DPC

Note

Go back to Example 21 and calculate the volume of excavation to form 1 m deep by 750 mm wide foundation trenches for that example.

Circles and triangles

In addition to those shapes of walls and buildings used as examples in this chapter so far, there are two others that the bricklayer will encounter in construction, which require measurement for setting out and calculation: the triangle and the circle, or parts of them (Figure 3.20).

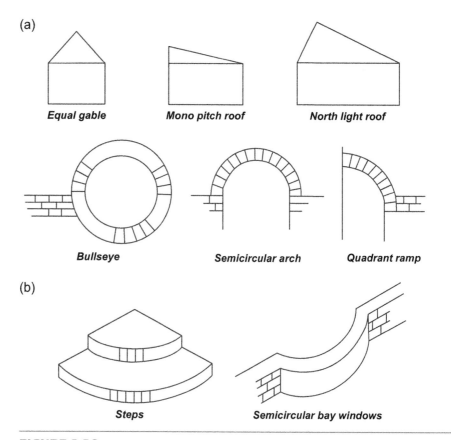

FIGURE 3.20

Examples of (a) triangular gable end walls; (b) circular shapes or parts of a circle used in construction

Example 30

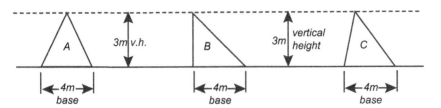

The surface area of any triangle can be calculated as follows, whether shaped like A, B or C:

Surface area = (Base × V ertical height) ÷ 2

= (4 m ×3 m) ÷ 2 = 6 m²

The surface area of all three triangles is 6 m², despite their different shapes.

Example 31

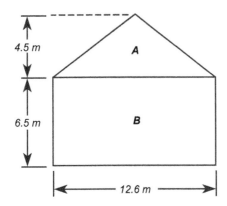

Calculate the surface area of the gable end wall, for repointing.

Answer 31

Total surface area = A + B

Area of A = (Base × Vertical height) ÷ 2

= (12.6 m × 4.5 m) ÷ 2

= 56.7 ÷ 2

= 28.35 m²

Area of B = 12.6 × 6.5

= 81.9 m²

Total surface area = 28.35 + 81.9 = 110.25 m²

The area of any circular shape can be calculated from the simple formula:

Area of circle = πr^2

(pronounced pi, r, squared).

Some pocket calculators have a button marked p; if not, then use the figures 3.14 (the value for pi never changes).

The letter r in the formula stands for radius of the circle.

The perimeter of a circle is called the circumference. Another simple formula is used to calculate the circumference of a circle:

Circumference of circle = π D

(pronounced pi, D). π remains 3.14 and D stands for the diameter.

Example 32

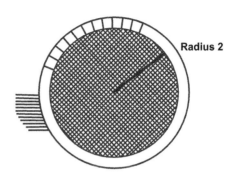

Radius 2

Calculate the surface area of the shaded part of the bullseye.

Answer 32

Surface area = π r^2

= π × r × r

= 3.14 × 2 × 2

= 12.56 m^2

Example 33

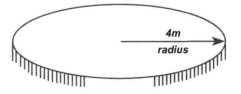

4m radius

Calculate the surface area of the circular paved area.

Answer 33

Surface area = π r^2

= π × r × r

= 3.14 × 4 × 4

= 50.24 m^2

Example 34

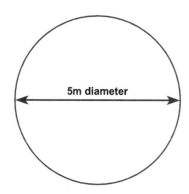

5m diameter

Calculate the circumference of the circle shown.

Answer 34

Circumference = π *D*

= π × Diameter

= 3.14 × 5

= 15.70 m

Estimating

Once the trainee has mastered the previous pages of calculations the knowledge gained can be used to calculate estimates for small building works.

An estimate is a price given for a certain amount of work according to the firm's set labour rates. These can change according to the firm.

On large contracts it is customary to have a bill of quantities. Under the traditional competitive tendering procedure the bill of quantities is the starting point for the main contractor tendering for the work.

The bill is drawn up by a quantity surveyor taking measurements from the final or contract drawings supplied by the architect.

From the quantities of materials shown in the bill and the nature of the finished building as shown by drawings and specifications, etc., the contractor estimates the costs for completing the contract.

The main areas included in the final estimate are:

- labour
- materials
- plant and equipment
- overheads.

The cost of labour is calculated by the application, to the total quantities, of labour rates per unit quantity, e.g. £15.50 per cubic metre for excavating, £12.60

> **Note**
>
> Remember that the circumference of a circle is just a plain measurement of length, in linear metres. Surface area is measured in m².

per cubic metre for clearing away, and so on through the trades, or by the estimated costs of complete operations such as those used for bonus targets.

The cost of the material comes from expected market prices plus any extra for handling, etc. Plant and other expenses may be allowed for in the labour rates or, preferably, estimated separately.

All these, when added together, give a total figure for the contract excluding overheads and profit. Overheads are items of expenditure that cannot easily be added to each contract, e.g. telephone, stationery and main office running costs. A percentage is added to all final figures to include these overheads.

The final percentage to be added is the profit required by the firm. The final figure plus overheads and profit gives the final estimate for the work.

Variations

This estimated price and the final price for a contract can change according to certain conditions.

The client could require extra works or prefer more expensive items than originally priced for, such as floor coverings.

The Building Regulations could change and cause the cost to increase.

If the contract architect made a mistake in the drawings this could lead to extra or less work.

Profits for the builder could be affected if there were mistakes in the calculations for the estimate. It is therefore essential that all calculations are checked and double checked. A simple mistake could cause a big loss.

Relaying information

To be able to relay information gathered to the appropriate people, good communication skills are required. Good communication skills are important in every aspect of our lives; in the workplace they are essential.

You may at times be the first point of contact with the client or customer, whether you are self-employed or working for a company.

Meeting and talking to people outside your organization can be one of the more interesting aspects of any job, but it is also a challenge.

You may be talking to people on the telephone or face to face, but in each case you are acting as the representative of your company and therefore you should present a positive and friendly image.

As firms grow into very large firms it becomes even more essential to have good communication.

Methods of communication

When entering the construction industry you will have to communicate well, irrespective of your position in the company.

It may be necessary to communicate with customers, visitors and potential employers in a number of different ways, for example:

- orally – face to face

- in writing

- by telephone

- through drawings.

Orally – face to face

When dealing with a customer or visitor face to face, always appear helpful and polite. Smile but do not appear to be over-friendly as this can make some people feel uncomfortable.

Never act in an aggressive or immature manner.

Remember, first impressions count. If your attitude comes across as 'couldn't care less', potential customers may think that this is your attitude towards your standard of work. Obviously this can frighten off potential customers.

Always offer as much advice and information as possible. If you cannot answer their questions or deal with their requests, find someone who can.

Verbal communication can include talking on the telephone, mobile or walkie-talkie. Verbal communication is instant and feedback can be obtained immediately.

Written

Written communication can take many forms and one or more of the following will commonly be used:

- letters

- memos

- reports

- records

- site diary

- handbooks and manufacturer's details

- e-mails and text.

The main advantage of written communication is the evidence it provides. Written communication may take longer than verbal communication, but with e-mails and mobile texting written communication is becoming much more rapid than in the past.

Letter writing

Writing a letter is an effective form of communication. We need to write letters on a number of occasions to:

- give information

- ask for details

- confirm things

- make arrangements.

Letters provide a permanent record of communication between organizations and individuals.

They can be handwritten, but formal business letters give a better impression of the organization if they are typed.

They should be written using simple, concise language. The tone should be polite and business like, even if it is a letter of complaint.

The letter must be clearly constructed, with each fresh point contained in a separate paragraph for easy understanding. When you write a letter for any reason, remember the following basic structure:

- your own address

- the recipient's address

- date

- greetings

- endings

- signature.

Report writing

A report is written to pass on information quickly and accurately to another party.

When writing a report, sufficient information should be included to allow the reader to understand fully what the report is intended to say.

A report should be divided into five parts:

- headings

- introduction to the report

- body of the report

- conclusion, with any recommendations

- any data, drawings, contracts, etc.

Site diary

This is one of the most important reports that a site manager has to produce during a contract.

It will include the following:

- the weather, morning and afternoon

- any visitors on site

- delays by subcontractors

- instructions from the clerk of works

- materials delivered
- number of staff on the site.

Telephone

Telephones have a very important role to play in communications. Often, a telephone conversation may have taken place before a letter is written to confirm something. Its clear advantage over the written message is obviously the speed with which people are put in touch with one another.

A good telephone manner is essential. Remember, the person you are talking to cannot see you so you will not be able to use facial expressions and body language to help make yourself understood.

The speed, tone and volume of your voice are very important. Speak very clearly at a speed and volume that people can understand. Always be cheerful and remember that someone may be trying to write down your message.

Making notes

It is important when receiving information that some kind of record is kept. Very often information is not written down and then mistakes are made through not remembering correctly.

Instructions are very often given by word of mouth. They should be kept brief and to the point, without complicated and confusing detail. This should make it easier for the recipient to remember them.

It is always good practice to aid your memory by making notes of what is being said. In order to record and possibly pass information received to others correctly, keep a small notebook in your pocket and always use it if the oral information is complicated and has to be passed on to others.

You may not be able to write down the instructions word for word.

- Make lists.
- Use notes and sketches.
- Abbreviate notes using key words.

Mon 22/01/09

11.45 am Mr Smith
12 Short Street

Repoint back of house

Needs estimate for the work

Positive and negative communications

For a company to function properly and make a profit, it is essential to maintain good working relationships within the workforce.

To achieve this, various sections and members of the workforce must communicate with one another as fully as possible.

When you first join a company you are often the odd one out and therefore you should try hard to co-operate with your new work colleagues. This will mean being polite, acting on request as quickly as possible and building a good working relationship with workmates.

You may be talking to people on the telephone or face to face, but in each case you must try to be both positive and friendly, both to give a good account of yourself and to promote good, effective working relationships.

Poor communications cause negative feedback and can lead to delays and mistakes which will cause a loss of profit for the company.

Barriers to communication

In all methods of communication there can be obstacles and these are known as barriers. These are of two kinds, external and internal, depending on their origin.

External barriers are those things in the working surroundings that interrupt or suppress the process of communication, such as machine noises, telephone bells, people talking and other similar distractions.

External barriers are usually quite easy to identify and overcome, whereas internal barriers, which arise within a person, are more difficult to recognize and combat.

Lack of knowledge and experience, different responsibilities, background and levels of intellect create problems for the communicator.

It is the emotional make-up of people, however, that gives the biggest headache. Their instinctive response to circumstances and to what others expect of them results in the formation of attitudes and behaviour that are often unexpected.

Principles of customer service

Firms

Customer relationship management is the process of establishing and maintaining a strong rapport between the organisation and its customer base as a means of building support or loyalty.

Modern construction firms have a web site that will attract new and existing customers.

Employee

As a new starter in the construction industry it is important to adopt a certain standard that will carry you forward in your new employment.

Most of the time you will be working on a building site or in a client's home. Always respect the property and treat it with care and attention.

Behaviour

You should always behave in a respectable manner that meets the requirements of the company and the client.

Professional image and appearance

Always dress the part. It is not a fashion show. Wear the appropriate PPEs provided by your employer. There may be times in a client's home that some items will not be required. For example fixing a fireplace inside a newly completed, or existing building, hard hats can be removed.

Personal hygiene

Always make sure your personal hygiene is adequate for the company and client. Think about who you are working with and where.

Use of appropriate language

Use appropriate language for the situation.

Good relations, establishing the needs of others

Always take care to respect other tradespeople you may be working with. You may at time have to help or assist other tradespeople.

Legal considerations

There may be paperwork which has to be completed in each day/week. Time sheets, travelling etc. ensure they are correctly completed and sent to the correct person.

Respect of the working environment

As mentioned previously always respect the environment you are working in. Ensure the area is kept neat and tidy and ensure all waste is correctly disposed of.

Good timekeeping

It is essential that good timekeeping is maintained especially if you have to catch transport to the working site. Your delay can then cause others to be delayed.

Produce a development plan

Personal development planning is the process of: establishing aims and objectives (or goals) - what you want to achieve or where you want to go, in the short, medium or long-term in your career.

Identify own development needs

- It is important that the apprentices / students can identify their own development needs.

- There should be many chances to increase knowledge and experience in this vast industry.

- It is also important for our employer to see your potential as a possible leader.

Producing a Personal Development Plan

- Set yourself goals.

- Prioritise those goals.

- Set yourself deadlines for when you want to achieve them.

- Recognise threats and opportunities.

- Develop your skills or increase your knowledge.

- Use your support network.

- Measure your progress.

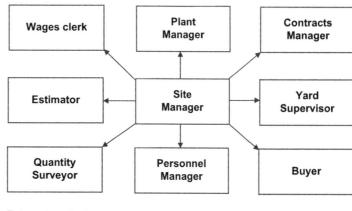

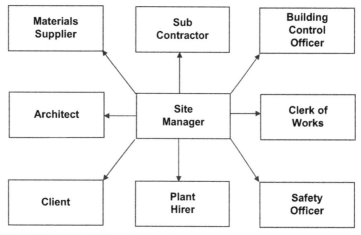

FIGURE 3.21
Internal and external contacts

Communications with those outside the building team

Examples of internal and external contacts are shown in Figure 3.21.

To be beneficial to both head office and the site manager there should be two-way communications to show how work is progressing, any problems that may have developed, and whether any more information, materials, plant or labour is required to maintain progress.

Multiple-choice questions

Self-assessment

This section of the book is designed to allow you to check your level of knowledge. The section consists of revision questions for this chapter. The questions are all multiple choice and have four possible answers. The answers are to be found at the end of the book.

The main type of multiple-choice question will be the four-option multiple-choice question. This will consist of a question or statement, known as the stem, followed by a choice of four different answers, called the responses. Only one of these responses is the correct answer; the others are incorrect and are known as distracters.

You should attempt to answer the questions by choosing either (a), (b), (c) or (d).

Example

The person employed by the local authority to ensure that the Building Regulations are observed is called the:

    (a) clerk of works

    (b) building control officer

    (c) council inspector

    (d) safety officer

The correct answer is the building control officer, and therefore (b) would be the correct response.

Communication

Question 1 An assembly drawing would be most likely to be drawn to a scale of:

    (a) 1:5000

    (b) 1:500

    (c) 1:50

    (d) 1:5

Question 2 Which of the following components are best suited for a schedule?

    (a) wall ties

    (b) windows

    (c) air bricks

    (d) insulation batts

Question 3 Which of the following gives a precise description of the materials and workmanship required for a contract?

    (a) specifications

    (b) schedules

    (c) schemes

    (d) contract plans

Question 4 Which of the following formulae can be used to calculate the area of the triangle shown?

(a) base × height
(b) base × slope ÷ 2
(c) base × height ÷ 2
(d) base × slope

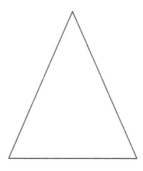

Question 5 Which of the following is the correct area for the rectangle shown?

(a) 207 m^2
(b) 32 m^2
(c) 64 m^2
(d) 20.7 m^2

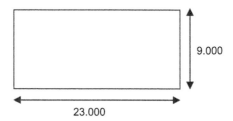

9.000

23.000

Question 6 Which of the following methods of communication would be written?

(a) radio
(b) letters
(c) mobile
(d) telephone

Question 7 Which method of projection is shown?

(a) oblique
(b) orthographic
(c) isometric
(d) cavalier

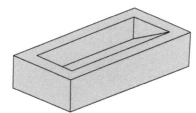

Question 8 Which of the following could result in the original estimate to the client being altered?

(a) rise in the cost of materials
(b) alteration to the programme
(c) rise in the cost of living
(d) alteration to the working week

CHAPTER *4*

Construction Technology

This chapter will cover the following City and Guilds units:

- Apprenticeship   Module 11   S1.1, S1.2, S2.1, S2.2, S2.3
- Certificate   Unit 205   1.1, 1.2, 1.3, 2.1, 2.2, 3.1, 3.2, 3.3, 3.1
                     Unit 201   2.1, 4.1, 4.2, 5.1, 5.2, 5.3, 5.4, 5.5
- Diploma   Unit 202-002   11.1, 11.2, 11.3, 11.4, 11.5
                                     12.1, 12.2, 12.3, 12.4, 12.5, 12.6
                                     13.1, 13.2, 13.3, 13.4, 13.5, 13.6, 13.7
                                     14.1, 14.2, 14.3
                                     15.1, 15.2
                     Unit 206   4.1, 4.2

This chapter is about:

- Interpreting building information
- Planning and carrying out productive, efficient working practices
- Understanding building substructure and superstructure
- Understanding construction materials and their environmental considerations

The following Skills and Knowledge requirements will be covered:

- Read and follow a working drawing to set out a building
- Transfer building line positions to the foundation concrete

The following Certificate learning outcomes will be covered:

- Understand how to prepare construction sites for setting procedures
- Set out rectangular shaped structures
- Know building substructure and superstructure

The following Diploma learning outcomes will be covered:

- Know about environmental considerations
- Understand construction foundations
- Understand construction of internal and external walls
- Understand the construction of floors, roofs

Working drawings

It has been stated before that drawings are the best way of communicating detailed information from the designer to all those involved in the works. For more details on working drawings see Chapter 3.

Because drawings are part of a legal contract between the client and designer it is essential that all drawings are precise and easy to interpret. Mistakes in either the drawing or the interpretation can be very costly.

All drawings are produced according to the guidelines produced by the British Standards Institute, BS 1192: Construction Drawing Practice. These standards ensure that all drawings are drawn to the same standard and should make it easier for everyone to interpret them in the same way.

Setting out

The person in charge of the site has to read and understand the drawings in order to be able to set out the building. It is essential that this part of the works is carried out correctly as mistakes in setting out can be costly.

Preliminary setting out of a building

Full setting-out procedures are covered in more detail in Chapter 6.

Before any setting out should be undertaken the top soil should be removed.

It is assumed that the building of a traditional semi-detached domestic dwelling is contemplated, but it should be noted that the same principles can be applied to all types of construction buildings and other more complicated forms (Figure 4.1).

When the setting-out stage is reached the site will have been previously surveyed and probably levelled to find the shape and features of the land.

Drawings showing the proposed buildings and associated site works will have been submitted to the local authority and approved.

A large development requires setting-out drawings to be prepared by the architect or engineer, but a small site would have sufficient information on the block plan or general site plan.

The most important task in setting out a building is to establish a base line to which all other setting out can be related.

This base line is very often the building line; this is an imaginary line that is established by the local authority.

You can build on the building line and any distance behind the building line, but

NEVER BUILD IN FRONT OF THE BUILDING LINE.

It is usual for the building line to be given as a distance from one of the following:

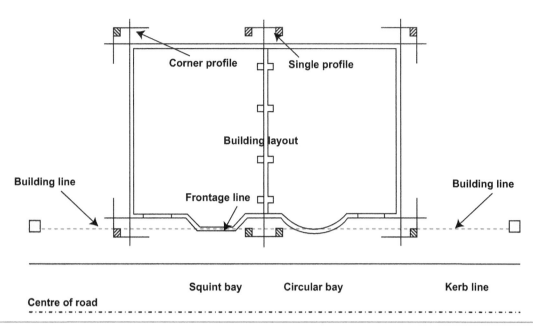

FIGURE 4.1
Location of foundation profiles

- existing buildings
- the centre line of the road
- the kerb line.

The frontage line of the proposed building must then be set out on or behind the building line, but never in front of it. In this example the frontage line is on the building line.

Degree of accuracy

A high degree of accuracy is required when setting out and this can be achieved if a steel tape is used and supported to avoid any sag.

Checks should be made on all measurements wherever possible.

All horizontal measurements should be made horizontal. Use a level to ensure that the tape is horizontal to avoid errors (Figure 4.2).

The building line is fixed to two pegs on either side of the proposed plot, a reasonable distance away from the building.

The two corner pegs are then set out on the building line, which marks out the frontage line of the proposed building (Figure 4.3).

The side walls of the building can now be set out at right angles to the frontage line. Ensure that the pegs are placed at a distance more than the side wall length (Figure 4.4).

The right angles can be set out with a builder's square, the 3:4:5 method or an optical square.

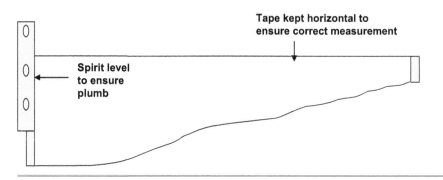

FIGURE 4.2
Accuracy with horizontal measurements

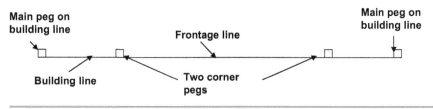

FIGURE 4.3
Setting out frontage line

Once the position of the side walls has been established the correct length can be measured and pegs can be knocked in. The back line can now be completed.

Once the outline has been completed it is essential that the setting out is checked for square. The two diagonals are measured and if they are equal the setting out is accurate and the building is square (Figure 4.4).

If they are not equal adjustments have to be made. Remember never to alter the frontage line as this should be correct.

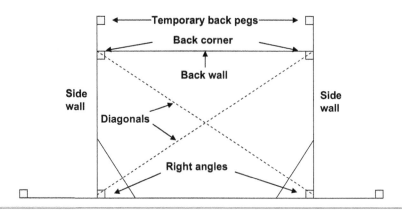

FIGURE 4.4
Checking for accuracy

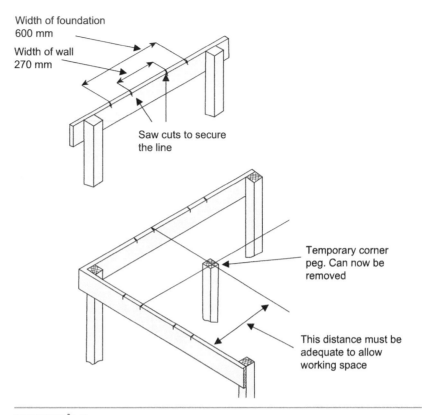

Width of foundation
600 mm

Width of wall
270 mm

Saw cuts to secure
the line

Temporary corner
peg. Can now be
removed

This distance must be
adequate to allow
working space

FIGURE 4.5
Types of profile

The back two pegs have to be adjusted to ensure that the diagonals are equal.

When adjustments have been made always recheck the dimensions of all wall lengths.

When the setting out has been proven correct the profiles have to be erected. At the moment the setting out pegs will be in the foundation trench so they have to be repositioned approximately 1 m away from all the wall faces. Profiles can be either corner or single type and consist of wooden pegs and rails (Figure 4.5).

Profiles are erected to enable the corner setting out pegs to be removed to allow excavation to take place without disturbing the pegs.

One of these corner profiles could be set to a given level which is known as the datum, and may relate to the finished floor level or damp-proof course (DPC). This datum peg should be protected with concrete.

The datum peg could also be positioned away from the setting pegs but close enough to be accessible and should also be protected with concrete and a small barrier of pegs and rails (Figure 4.6).

The profiles should be positioned approximately 1 m away from the face of the building to allow working space for the excavation.

The completed profiles can have foundation and wall widths marked on them

FIGURE 4.6
Datum peg

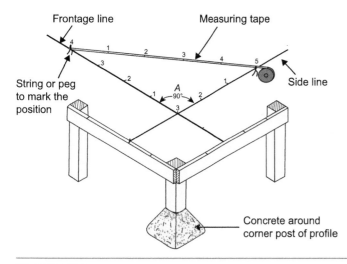

FIGURE 4.7
Checking a right angle by the 3:4:5 method

in one of various methods. Saw cuts are best as nails could be accidentally removed (Figure 4.7).

Once the profiles have been accurately constructed the dimensions should all be checked again, as should both diagonals.

The original setting-out pegs can now be removed.

Building lines can now be fastened to the profiles and the trench marked out ready for excavation (Figure 4.8).

Sand can be used to mark out the ground ready for excavation and the lines can then be removed.

Ground works

As soon as the setting out is completed the excavation can begin

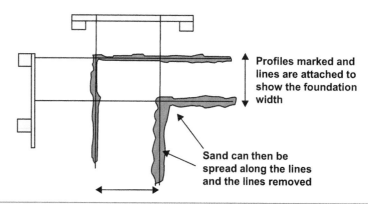

Profiles marked and lines are attached to show the foundation width

Sand can then be spread along the lines and the lines removed

FIGURE 4.8
Marking out the ground ready for excavation

Excavations

On small sites, hand-held tools such as picks, shovels and wheelbarrows are used. However, if the depth of excavation exceeds 1.2 m some other method of removing spoil from the excavation will have to be used.

On all sites mechanical methods could be used, depending on factors that are different on each site. These include the volume of soil involved, the nature of the site and any cost constraints.

Before excavation begins it is essential to check for underground cables and services and ensure that they are adequately marked and disconnected if necessary. Cable and pipe detectors are available to detect buried services.

There are national colour codes for buried services, the most common of which are:

- black – electricity
- red – high-voltage electricity
- blue – water
- yellow – gas
- grey – telecommunications
- green – cable television.

It is now common for coloured polythene tape to be placed in the backfill about 300 mm above the pipe or service cable.

The most common machine for digging site trenches is the backactor (Figure 4.9).

Work below ground level

Almost all construction sites will involve some form of excavation. Every year many construction workers are involved in accidents connected to excavations.

FIGURE 4.9
Backactor

The most common problem is the collapse of trench sides or equipment or materials falling into trenches.

Any ground work must be planned and carried out correctly and safely to prevent accidents.

All subsoils vary in their abilities to remain stable during excavations.

The first operation is the removal of the top soil, which should have been completed before the building was set out. This also makes the marking out of the ground easier.

Top soils can vary in depth from approximately 150 mm to 300 mm, and they contain various organic materials, making them unsuitable to build on. They could be either removed from the site or stored for later use when landscaping.

The foundation trenches can now be excavated by either hand or machine. Hand excavation is only carried out where the site restricts the use of machinery, such as for small extensions at the rear of existing properties.

Timbering

All subsoils have a natural angle of repose which allows the subsoil to rest unless given support. As the trench is excavated temporary support is placed in the trench according to the type of subsoil.

Some simple forms for shallow trench excavations are shown in Figure 4.10.

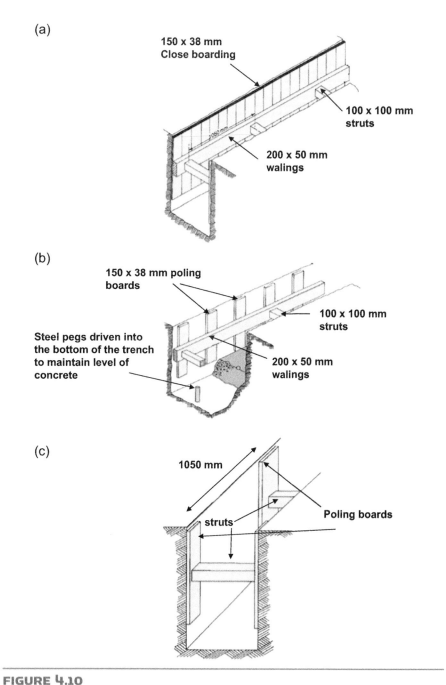

FIGURE 4.10
(a) Loose soil; (b) moderately firm soil; (c) firm soil

Laying concrete foundation

After the trenches have been excavated to the required depth the concrete can be placed to form the foundation. A straight edge should be used between each steel peg to produce a level foundation on which to build (Figure 4.11).

A rough finish is required to ensure that the brickwork adheres to the concrete foundation.

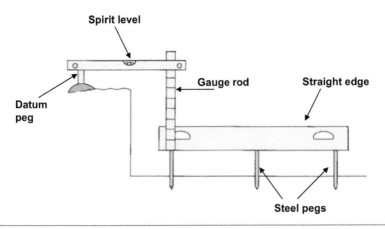

FIGURE 4.11
Placing concrete in foundations

The distance from the top of the datum peg to the concrete needs to be in brick courses. This is checked with a spirit level and gauge rod (Figure 4.11).

The pegs are then levelled around the trench with either a straight edge or a spirit level. On larger areas, or where more accuracy is required, there are numerous mechanical levels available such as the quickset level, the autoset level and the laser level.

Brickwork up to damp-proof-course

Once the concrete has hardened the brickwork can be built up to DPC level.

Lines are fastened to the profiles to mark out the position of the brickwork (Figure 4.12).

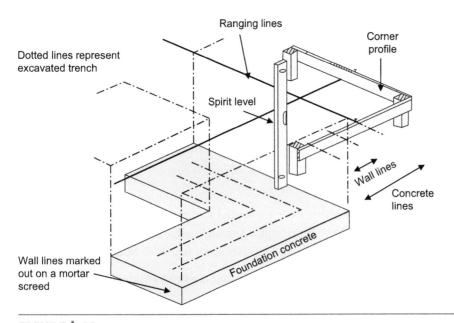

FIGURE 4.12
Plumbing down from ranging lines

The ranging lines are set to wall marks on the profiles, and plumb lines are taken down on the main corners to the concrete foundation.

To facilitate clear marking, a mortar screed must be spread over the concrete to a thickness of approximately 3 mm at the positions where the plumb lines are being taken.

A spirit level can normally be used to ascertain plumb lines, but if the trench is too deep to allow this to be done, a 'drop bob' should be used (Figure 4.12).

The bricks are laid and the corners are erected by racking back until the correct height is reached.

The first course should be dry bonded to ensure that the correct bonding pattern is maintained.

The top two courses below the DPC should be facing bricks as they will be seen after the ground work has been completed.

Once the corners have been erected the main walls can be completed using lines and pins.

The procedure for working below ground level is very different to working above. The space in the trench may be very tight and all materials have to be set out on the sides of the trenches. Working below ground is often known as nose bleeding as you are mainly bending over.

It is important not to place the materials too near the side of the trenches, which may cause collapse due to too much pressure (Figure 4.13).

If the foundation brickwork is a cavity wall there will be more space for the feet while erecting the face wall. Solid brick or foundation block walls will be tight for space as most foundations are designed with a minimum 150 mm space on either side of the walls.

At this stage it is sufficient to know that the ground-floor level is approximately 225 mm above DPC level.

When this point is reached in the building of the wall, a level must be transferred from the datum which represents ground-floor level. This level will have

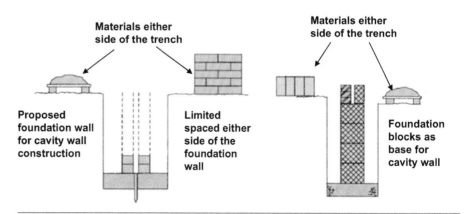

FIGURE 4.13
Preparation for work below ground level

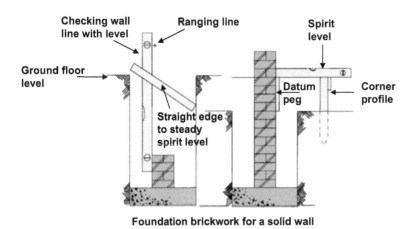

Foundation brickwork for a solid wall

FIGURE 4.14
Transferring setting out lines and datum to substructure

been adhered to while constructing the wall, to ensure that the wall is horizontal at all points.

The transferred level is maintained on the wall by fixing a short length of 50 mm × 25 mm batten, termed a datum peg (Figure 4.14).

From this datum a storey rod is used to maintain correct heights and, to be of assistance to the charge hand bricklayer, it should be marked with all those details appearing on the elevation of the wall, e.g. sills, arch and floor levels (Figure 4.15).

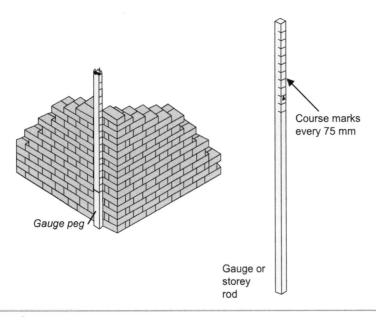

FIGURE 4.15
Use of storey rod

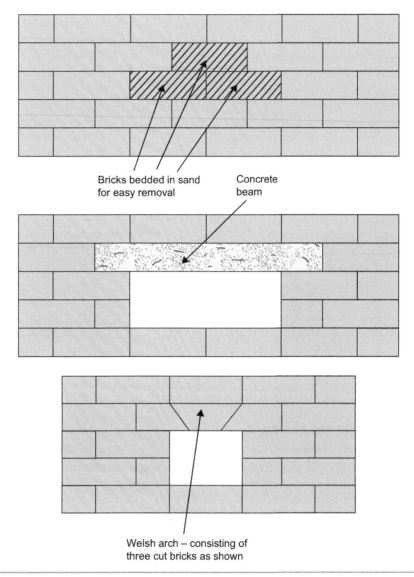

Bricks bedded in sand
for easy removal

Concrete
beam

Welsh arch – consisting of
three cut bricks as shown

FIGURE 4.16
Provision for services

Services

During the work below ground level there are times when openings have to be left in the brickwork for services, etc. (Figure 4.16).

Sometimes the exact position is not known, so sand courses are used to allow bricks or blocks to be removed at a later date.

Foundations

The foundation of a building is that part which is in direct contact with the ground.

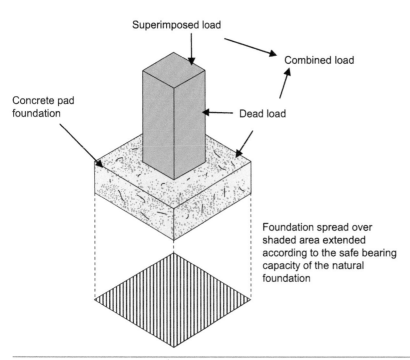

FIGURE 4.17
Loads

The current Building Regulations require that the foundations of a building shall safely sustain and transmit to the ground the combined dead and superimposed loads in such a way as not to impair the stability or cause damage to any part of the building.

The ground or subsoil on which a building rests is called the natural foundation or subfoundation, and has a definite load-bearing capacity, according to the nature of the soil (Figure 4.17).

Subsoils are of many varieties and may generally be classified as rock, compact gravel or sand, firm clay and firm sandy clay, silty sand and loose clayey sand.

It is possible to erect a wall on rock with little or no preparation, but on all soils it is necessary to place a continuous layer of in situ concrete in the trench, called the building foundation.

The term 'in situ' means cast in place in its permanent position; unlike 'precast' which is made elsewhere, lifted and transported later to the place where it is required for use.

This cast, in situ concrete is made from Portland cement with coarse aggregate, plus sharp sand or ballast, graded from 40 mm to fine sand, and mixed in the proportion of 1:6.

Foundation types

This chapter will deal with four main types of foundations:

- strip foundations – narrow strip and wide strip foundations
- pad foundations
- raft foundations
- pile foundations.

Strip foundations

When a load is placed on soil it is necessary to 'spread' or 'extend' the foundation base to ensure stability. This extended or spread foundation is referred to as a strip foundation in the case of a continuous wall structure, or a pad foundation in the case of an isolated pier.

The thickness of the concrete foundation should not be less than the projection of the strip either side of the wall, but in no case less than 150 mm. One method of ascertaining the depth of concrete is shown in Figure 4.18.

Atmospheric depth

This is the depth below ground level to which foundations should be taken. It depends on the type of soil and is the depth at which the subfoundation ceases to be affected by the weather. This is between 600 and 1500 mm, decreasing as the proportion of gravel increases (Figures 4.18 and 4.19).

Narrow strip

Narrow strip, or trench fill as it is also known, is used as an economical alternative to the normal strip foundation (Figure 4.19).

A narrow strip is excavated by the mechanical excavator and backfilled with mass concrete up to a level just below the finished ground level.

A high standard of accuracy in constructing such a foundation is required. It is

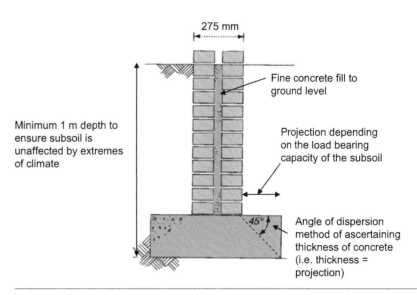

FIGURE 4.18
Strip foundation

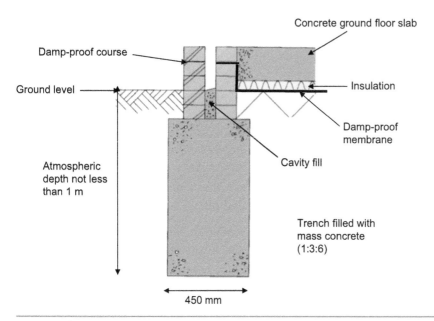

FIGURE 4.19
Narrow strip foundation

cheaper and quicker to fill the trenches with mass concrete than to excavate a wider trench. There is less excavated material to be removed, no brickwork below ground and backfilling is eliminated. This method also eliminates timbering to trenches.

Wide strip foundations

Where the structural loads are very heavy or the safe bearing capacity of the soil is low, the spread of the foundation base becomes greater. This is normally referred to as a wide strip foundation. It follows that the required depth of the concrete foundation base may be considered excessive and it can be reduced by the introduction of steel reinforcement, but the foundation must always be of sufficient depth to ensure that, in combination with the steel, it will resist the stresses of tension and shear. Figure 4.20 shows a simple example of a re-inforced concrete strip foundation.

Pad foundations

For single loads that are transmitted down a column, the most common foundation is a square or rectangular block of concrete of uniform thickness known as a pad foundation (Figure 4.21).

Raft foundations

These foundations consist of a raft of reinforced concrete under the whole of the building.

Raft foundations are often used on poor subsoils for lightly loaded buildings and are considered capable of accommodating small settlements of the subsoil.

> **Note**
>
> In order to spread the load over greater area it is necessary eithe to make the pad thicker or to us reinforced concrete.

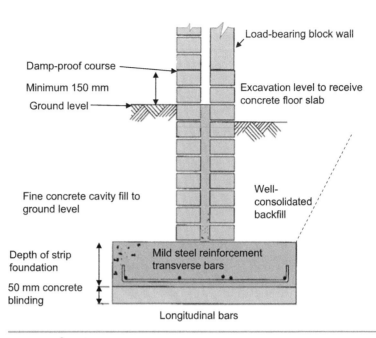

Load-bearing block wall

Damp-proof course

Minimum 150 mm

Ground level

Excavation level to receive concrete floor slab

Fine concrete cavity fill to ground level

Well-consolidated backfill

Depth of strip foundation

Mild steel reinforcement transverse bars

50 mm concrete blinding

Longitudinal bars

FIGURE 4.20
Reinforced wide strip foundation

The simplest and cheapest form of raft is the thick reinforced concrete raft (Figure 4.22). Its rigidity enables it to minimize the effects of differential settlement.

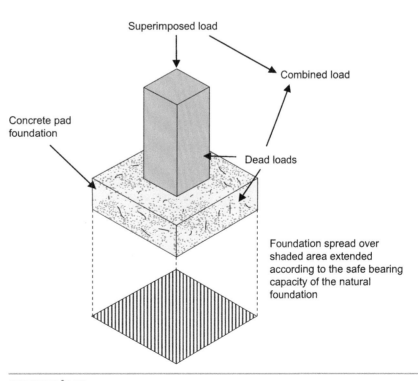

Superimposed load

Combined load

Concrete pad foundation

Dead loads

Foundation spread over shaded area extended according to the safe bearing capacity of the natural foundation

FIGURE 4.21
Pad foundation

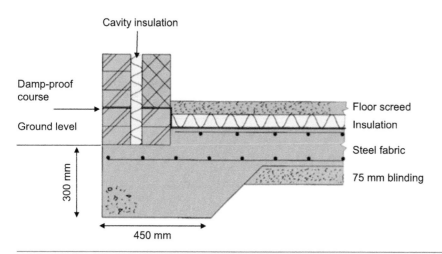

FIGURE 4.22
Raft foundation

Short bored piled foundations

If, instead of spreading the load from the wall over a wide area, it is decided to transfer it to a greater depth, an economical solution is the use of a short bored pile foundation.

Short bored piles are formed by boring circular holes 300 mm diameter to a depth of about 3 m by means of an auger. This depth is governed by the level of suitable bearing capacity ground. These are filled as soon as possible with mass concrete.

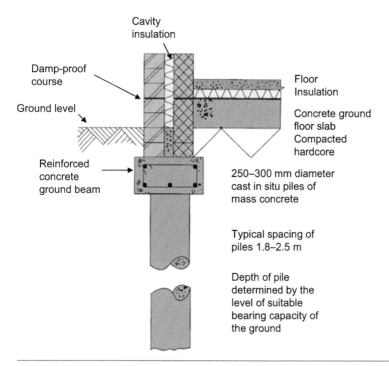

FIGURE 4.23
Piled foundation

The piles are placed at the corners of the building and at intermediate positions along the walls. The piles support reinforced concrete beams which are cast in place in the ground (Figure 4.23).

Floors

Ground floors

There are two types of ground floor construction found in both domestic and communal buildings:

- solid

- hollow.

The functions of ground floors are to provide a level surface with sufficient strength to support all the loads, and to prevent dampness entering the building and heat loss from the building.

Solid ground floors

Solid ground floor consists of a compacted hardcore with sand blinding on top. This protects the damp-proof membrane (DPM). Insulation is laid on the membrane on which a slab on concrete is laid.

Details are shown in Figure 4.24.

The horizontal DPC level must never be above the floor level.

Brick rubble or hardcore laid directly beneath the concrete floor will not only prevent settlement but, being of a porous nature, also help to prevent dampness. Its thickness should be approximately that of the concrete floor.

Solid ground floors have to be insulated, according to the current Building Regulations, to provide resistance to unacceptable heat loss through the floor.

This can be achieved in various ways but the most common is to place the insulation on top of the DPM, which is placed on a blinding layer on top of the hardcore.

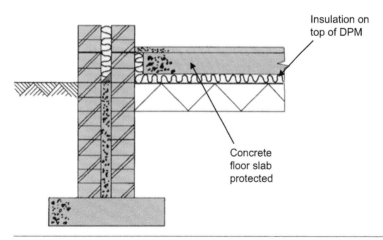

FIGURE 4.24

Damp proofing a solid ground-floor slab with the DPM under the concrete floor slab

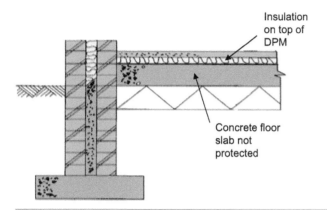

Insulation
on top of
DPM

Concrete floor
slab not
protected

FIGURE 4.25
An alternative method with the DPM on top of the concrete floor slab

This method of construction protects the concrete floor slab from any moisture or harmful salts. The only problem with this method is the risk of damage to the DPM and insulation when laying the concrete floor slab.

An alternative method is shown in Figure 4.25. This method is easier but the concrete floor slab is not protected against the ingress of moisture or harmful salts.

Hollow ground floors

Suspended timber floors need to have a well-ventilated space beneath the floor construction to prevent the moisture content of the timber rising above an unacceptable level, i.e. above 20 per cent, which would create the conditions for possible fungal attack.

Hardcore and oversite concrete are still required for these floors, but in this case the concrete does not require a waterproof membrane.

Hollow sleeper walls are constructed on the oversite concrete to receive the wall plate, which in turns supports the floor joists (Figure 4.26). A horizontal DPC is inserted under the wall plate to resist rising damp.

Underfloor ventilation against dry rot

Having placed the oversite, the floor must be supported, so as to allow free passage of air to prevent the floor timbers from rotting. This is achieved by building honeycomb sleeper walls on the oversite concrete and air bricks built into the external wall.

Two types are illustrated in Figure 4.27.

Precast beam and pots

Prestressed precast concrete beams were first used in commercial ground floor construction and are now common in domestic construction. They have been designed as an alternative to suspended timber floors. The same preparation is

> **Remember**
>
> The DPM must be continuous wi the horizontal DPC in the extern and internal walls.
>
> Materials suitable for floor insul. tion are dense resin bonde mineral or glass fibre slabs, ar polystyrene and cork slabs. Thes should be placed above the DP and turned up at the edges of th floor slab to prevent heat los through the external wall.

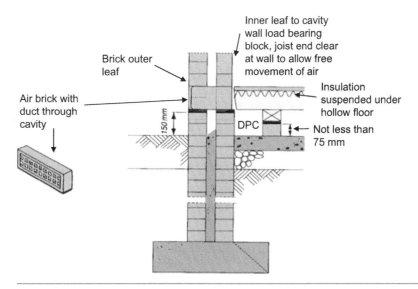

Inner leaf to cavity wall load bearing block, joist end clear at wall to allow free movement of air

Brick outer leaf

Air brick with duct through cavity

Insulation suspended under hollow floor

150 mm

DPC

Not less than 75 mm

FIGURE 4.26
Cavity wall, showing underfloor ventilation

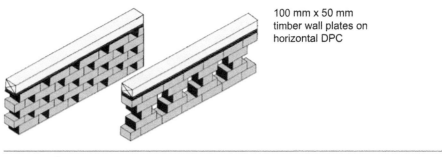

100 mm x 50 mm timber wall plates on horizontal DPC

FIGURE 4.27
Types of sleeper wall construction

required as for suspended timber floors (Figure 4.28). They consist of pre-stressed beams with block inserts.

When precast floors have been constructed they provide a safer work area than suspended timber floors at both ground and first floor level.

Suspended timber floors require boarding out before work can proceed, whereas the complete floor area of precast floors is ready to work on and can take full loads.

Floor finishes

There are numerous floor finishes available, some of which form an integral part of the construction, such as floor boarding and screeds.

Screeds

These are used to give the concrete floor a durable finish to receive the final floor finish, such as tiles.

FIGURE 4.28
Precast concrete ground floors

FIGURE 4.29
Cement and sand screeds to concrete floors

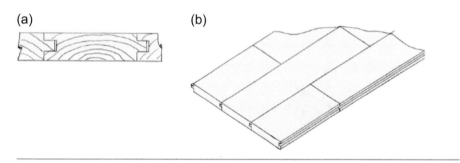

(a)   (b)

FIGURE 4.30
Covering to timber joists: (a) tongue-and-groove fboarding; (b) tongue-and-groove sheeting

Cement and fine aggregate screed are laid up to a thickness of 75 mm directly on top of the concrete floor (Figure 4.29).

Floor boarding

Timber suspended floors can be covered with either floor boarding or sheets (Figure 4.30).

Tongue-and-groove boarding is laid at right angles to the floor joists and should be fastened using floor brads.

Sheeting is available in 600 mm × 19 mm sheets and is laid at right angles to the floor joists in a chequerboard design. Sheeting should be screwed into position to prevent movement.

Floating floors

When quiet floors are required, such as for a library, floating floors can be constructed (Figure 4.31). They require the final floor to be separated from the floor structure to prevent the passage of sound up through the structure and into the room above.

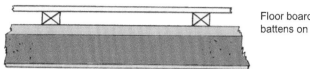

FIGURE 4.31
Floating floor

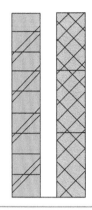

FIGURE 4.32
Cavity external wall

(a)   (b)   (c)

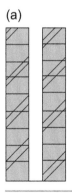 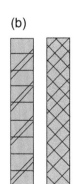 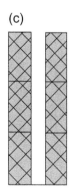

FIGURE 4.33
Types of cavity walling: (a) brick to brick; (b) brick to block; (c) block to block

Walls

External wall construction can consist of either solid or cavity walling (Figure 4.32).

Cavity walling can be further divided into masonry cavity walls (Figure 4.33) and timber-frame cavity walls (Figures 4.34 and 4.35).

Masonry cavity walls can be constructed in:

- internal and external skins of brick

- internal skin of block and external skin of brick

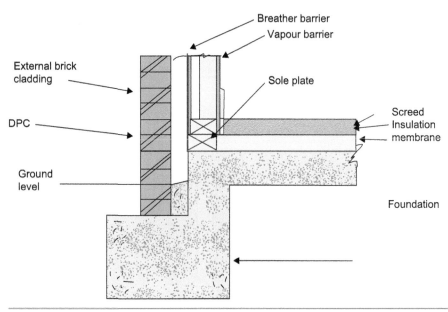

FIGURE 4.34
Timber frame

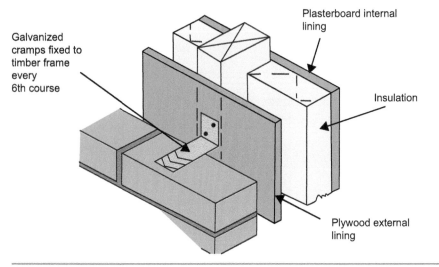

FIGURE 4.35
Timber frame fixing

- internal and external skins of blocks
- cladded walls.

External walls to framed structures may be grouped according to the type of structural frame: steel, concrete or timber.

This chapter will deal only with cladding to timber frames.

Timber frames

Timber wall frames are erected on a foundation of conventional walls or a concrete base or raft to which a sole plate is bolted.

The frame is usually assembled as a storey-height frame, in sections around the building.

Openings for windows and doors are included in the frame and are covered with a sheathing of timber boards or plywood and a breather barrier.

The brickwork is fastened to the timber frame with galvanized steel cramps fixed every sixth course.

The standard form of construction for the external walls of domestic and communal brick buildings is called cavity walling.

This means that the bricklayer builds the two separate leaves or 'skins' of brick masonry (a general term indicating brickwork and/or blockwork), with a 50–75 mm wide space between them.

The outer skin is usually 102.5 mm thick face brickwork, but may be constructed from facing quality blocks.

The inner skin is usually 100 mm thick common blocks that are later plastered to receive internal decoration.

Both skins of brick masonry are joined together with a regular pattern of corrosion-resistant ties, so that they behave as a single wall.

The main objectives of cavity walls are to prevent rain penetration and provide a greater degree of thermal insulation than solid wall construction.

New developments in the design of cavity walls have allowed for greater insulation values to prevent heat loss through the walls.

The external walls of domestic and communal buildings can also be designed in a framed construction. Frames can be timber, steel or concrete.

One of the great advantages of framed buildings is that the main construction can be almost finished and watertight before the external skin of brickwork is added. This is often considered as a faster method of building and allows other trades to continue while the external cladding is being fixed.

When designing to meet the current Building Regulations for heat loss through walls, it is important to consider which method of construction will be used, as the cost can vary according to which method is used.

The current Building Regulations require that external walls have a maximum U-value of 0.18 W/m^2 K. The U-value denotes the thermal transmittance, which is the rate of heat transfer through a wall from air to air.

The most common method of insulating cavity walls, whether they are brick and block or framed, is by inserting slabs of polyurethane foam, expanded polystyrene or bonded glass fibre into the cavity space.

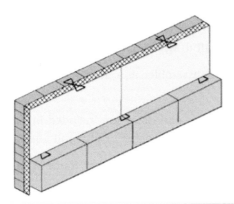

FIGURE 4.36
Insulation to cavity wall

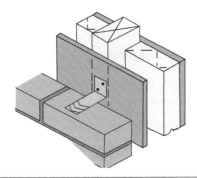

FIGURE 4.37
Insulation to timber frame cladding

Figures 4.36 and 4.37 show insulation placed in the cavity of a brick and block wall and insulation to timber-framed construction, respectively.

External envelope

The structure of every building is divided into two parts:

- superstructure – all elements of the building above ground level (horizontal DPC)
- substructure – all elements of the building below ground level.

This envelope can be described as the external envelope and is generally a combination of walls and roofs with openings for light and access.

External walls

The external envelope should provide the building with a watertight coat which is also decorative.

Facing bricks are the most common material used in modern domestic construction, whether the design is timber framed or brick and block cavity walling.

In certain parts of the country the local material will change and other materials will be considered, such as stone.

Whatever materials or methods are chosen for the external wall they all have to satisfy the following functions:

- strength and stability
- weather exclusion
- thermal insulation
- sound insulation
- durability
- fire resistance
- appearance
- access and egress.

Strength and stability

The external walls have to be strong enough to carry the loads without excessive deformation and to transfer these safely to the structural frame or foundation.

Weather exclusion

Keeping out the weather is an extremely important consideration if the internal environment is to remain constant.

Thermal insulation

Heat will flow from a high temperature inside the building to a low temperature on the outside. The external walls should be designed to reduce this heat flow to a minimum.

During the winter period there will be a constant flow of heat from the building which means that extra heating will be required to counteract the heat loss through the external envelope.

Sound insulation

Noise is defined as 'unwanted sound'.

Extremely loud noises can be detrimental to the health of the users of the building, while small noises can be irritating and could cause loss of concentration.

Since the majority of people do not have control of the sounds outside the building, such as road traffic and aircraft noise, the external envelope must reduce that noise to an acceptable level.

Durability

The materials from which the external walls are constructed must have suffi-cient resistance to the damaging effects of the climate – in the form of erosion, atmospheric pollution, frost, rain, and chemical and solar degradation – to provide a building that will be relatively maintenance free for its anticipated useful life.

Fire resistance

The building must be able to contain any fire until the fire-fighting appliance arrives.

The external envelope should be able to contain the fire and to prevent the spread of fire to other buildings in the vicinity.

Appearance

This is the aesthetic requirement which should be considered at the design stage.

The external envelope should be compatible with others in the vicinity, by either blending in with them or providing a contrast which is not too stark.

Remember that in the rural areas the envelope may be required to blend in with the landscape so that it becomes as inconspicuous as possible.

Access and egress

The external envelope has to have openings formed to allow the users in and out of the building, and to allow light to enter.

Internal walls

These usually act as partitions or dividers for the internal space of the building, although occasionally they also support the upper floor and roof loads.

The most common types of internal walls (Figure 4.38) are:

- masonry partitions
- timber-frame or stud partitions
- demountable partitions
- proprietary systems (paramount).

Masonry partitions

These can be built in either lightweight concrete or clay blocks (Figure 4.39).

Masonry partitions are one of the strongest types of internal wall and will support the loads subjected by the floors, roofs and superimposed loadings from the inhabitants and furniture.

They are built exactly as external walls, without the finish to the joints that external walls require.

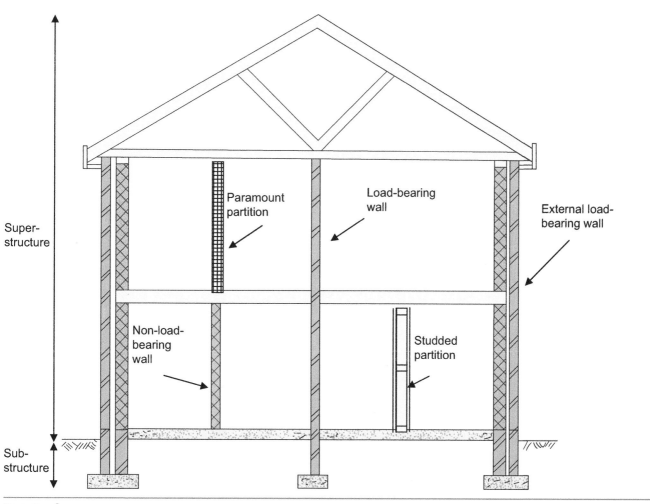

FIGURE 4.38
Types of wall

The mortar is usually weaker than the material being used, to allow any shrinkage to take place in the joint rather than in the block.

If they are built on wooden upper floors they require a sole plate. If they are built on a concrete ground floor a sole plate is not necessary (Figure 4.40).

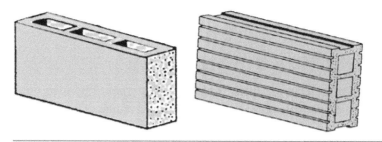

FIGURE 4.39
Types of lightweight internal block

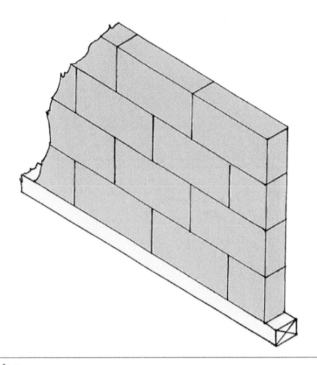

FIGURE 4.40
Lightweight block wall

Timber-frame or stud partitions

Timber studding is a non-load-bearing partition.

These partitions must be designed and constructed to carry their own weight and any fittings that may be attached to them.

They must be strong enough to withstand impact loadings on their faces and also any vibrations set up by doors being closed or slammed.

Details are shown in Figure 4.41.

Demountable partitions

A wide range of lightweight, non-load-bearing and demountable partitions is available (Figure 4.42). The construction is usually of metal, wood or plastic sections or trim with various infill panels in a variety of materials.

Their main advantages are their versatility and non-permanent nature. They can be used to alter the floor layout of existing properties with the minimum of disruption.

Proprietary systems (Paramount)

One particular proprietary partition consists of plasterboard bonded on either side of a strong cardboard cellular core to form rigid panels. These are fixed to ceiling and wall battens and supported on a timber wall sole plate.

These units can also be provided with suitable facing for direct decoration or even with self-finish applied (Figure 4.43).

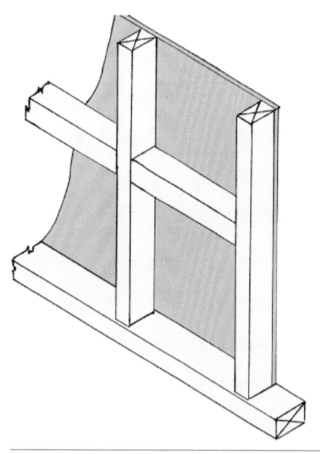

FIGURE 4.41
Timber studding

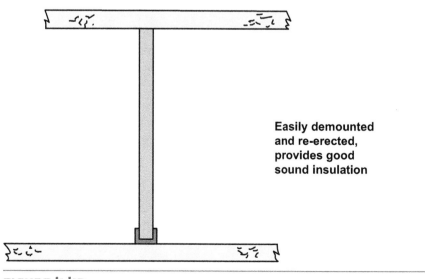

**Easily demounted
and re-erected,
provides good
sound insulation**

FIGURE 4.42
Demountable partition

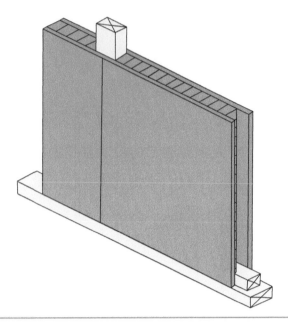

FIGURE 4.43
Paramount partitions

Roofs

The roof is that part of the external envelope which spans the external walls at their highest level and, being part of the envelope, it must fulfil the same functions.

Basic roof forms

These may be either flat or pitched and are named according to their shape (Figure 4.44). Any roof having a sloping surface more than 10 degrees is known as a pitched roof. A roof with a slope of less than 10 degrees is known as a flat roof.

A gabled roof has a double pitched roof between two triangular brick walls known as gables. These have soffit and fascia on the front and rear only.

A hipped roof is also a double pitched roof, but the eaves continue all the way around the building.

A lean-to consists of a single pitched roof resting against a building.

Roof components

The slope of the roof is known as the pitch. The edges of the roof are known as the eaves along the lowest horizontal edges and are terminated with a soffit and fascia (Figure 4.45).

The verge is found at the side and sloping edges and is finished with a barge board and soffit.

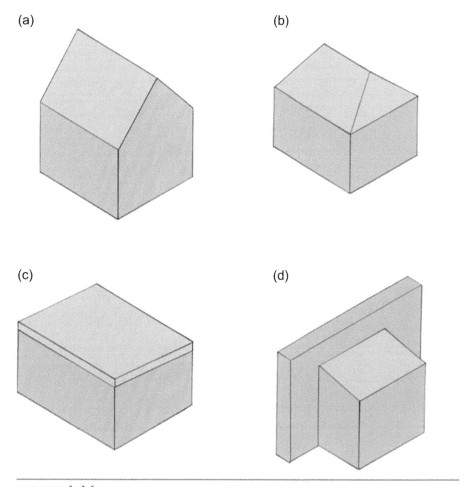

FIGURE 4.44
Basic roof forms: (a) gabled; (b) hipped; (c) flat; (d) lean-to

The detail of the eaves in Figure 4.45 shows tiles fixed to battens on sarking felt. This enables the roof to be watertight.

The sarking felt is continuous into the guttering fixed to the fascia.

The soffit has ventilation holes to allow the roof to be ventilated.

The roof is protected with insulation fitted between the ceiling joists.

Figure 4.46 shows roof details at the ridge. This detail shows slates fixed to battens on sarking felt on either side of the ridge of the roof.

Flat roofs

Flat roofs on domestic buildings are mainly of timber construction, while concrete flat roofs are used for industrial buildings.

Roofs with less than 10 degree slope are termed flat roofs. Most flat roofs have sufficient slope to prevent water standing on the roof surface.

Typical details of flat roofs are shown in Figures 4.47 and 4.48.

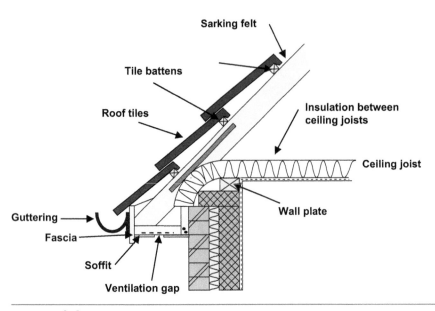

FIGURE 4.45
Basic roof terminology at eaves

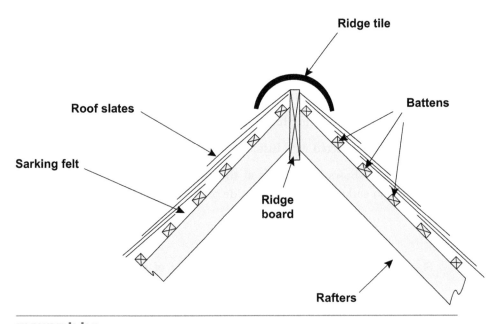

FIGURE 4.46
Basic roof terminology at ridge

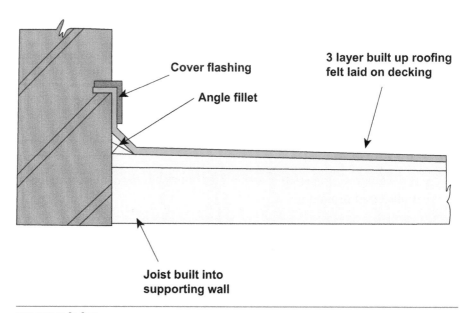

FIGURE 4.47
Junction of flat roof with main building

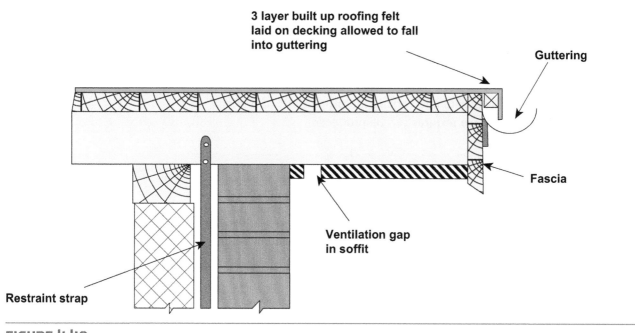

FIGURE 4.48
Flat roof eaves detail

Where the flat roof meets the main building precautions are taken to prevent water penetrating the building below.

The roofing felt is turned into the brickwork and cover flashings are placed over it to complete the seal.

At the eaves the roofing felt is allowed to fall into the guttering.

Materials

A wide range of materials can be used internally and these can be either manufactured or natural.

Properties of materials

All materials used in construction must:

- be of suitable quality
- be suitable for their intended use
- be readily available at an acceptable cost
- comply with the current relevant British Standards.

Aggregates

Aggregates are granular materials that are bonded together with a binder to produce a solid mass when set.

Aggregates can be either gravel or crushed rock and are divided into fine and coarse according to the size.

By using different types of aggregate and varying the proportions of fine and coarse aggregates, various surface textures can be produced.

Cement

Cement is a mixture of clay, gypsum and either chalk or limestone.

The materials are burnt and ground into a fine powder. When water is added to cement a chemical reaction takes place, which results in the cement setting and hardening, producing great strength.

Cement is the most common binder for concrete and mortar.

There are several types of cement, but the most common is ordinary Portland cement.

Concrete

Concrete is an artificial material made from a mixture of aggregates, cement and water.

Concrete is a very versatile material and can be moulded to fit any shape. Its strength can be increased by inserting steel reinforcing bars.

Lime

Lime has been used very successfully for thousands of years as the only cementitious ingredient in mortars.

The lime used then was hydraulic, meaning that it was made from a naturally occurring chalk raw material which contains clay impurities. This clay content gave the lime a slow setting action not unlike the chemical hydration-setting-hardening process of ordinary Portland cement, but the lime mortar would take years to harden.

The raw materials used to produce building lime are either chalk or limestone. When chalk or limestone is burnt at a very high temperature it turns into quicklime. This material cannot be used for building work in this state so it undergoes treatment by the addition of water, in a process known as slaking. The end product is hydrated lime.

Mortars

It is not usually difficult to cause an argument between bricklayers – just ask them the following question: Does mortar stick bricks together or keep them apart? The answer is of course that it does both, but a lot more besides.

Mortar must stick firmly to bricks and blocks in external walling to keep the rain out. Mortar bed joints also hold bricks apart, so that the courses can be kept level and to an even vertical gauge of four courses to 300 mm, with standard metric bricks.

When walls were much thicker and Portland cement had not been invented, a mortar mix of lime and sand was used very successfully for thousands of years. The lime used then was hydraulic, as described above, and the lime mortar would take years to harden.

Purpose

Bricklaying mortar is the ideal material for getting bricks to rest firmly upon each other, whether these are accurately shaped class A clay engineering and calcium silicate bricks or more irregularly shaped hand-made bricks.

The mortar must remain soft enough for each brick to be pressed down to the line, before suction causes the bed to stiffen up. Not only does mortar accommodate irregularities, but it must stick firmly to each brick so as to stop rain from penetrating exposed joints.

Bricks

The study of bricks, from raw materials to delivery of finished products, is an extensive one.

Being able to recognize a brick when it appears on site, know its properties such as shape, size, weight, strength, porosity and colour – and therefore know how and where to use it correctly – is all important basic knowledge.

Brick making is a very skilful business, with many individual variations in methods of manufacture between companies and their factories.

British Standards specify a brick as a walling unit designed to be laid in mortar and not more than 337.5 mm long, 225.0 mm wide and 112.5 mm high, as distinct from a building block which is explained as a unit having one or more of these dimensions larger than those quoted for bricks.

Bricks, which are one of the most durable materials, can be described as building units which are easily handled with one hand.

There are numerous uses for bricks but the main ones are as units laid in mortar to form walls and piers and, increasingly, for brick paving.

Bricks were first made many thousands of years ago in hot climates, where a clay mixture was moulded and dried in the sun. It was found that if the clay mixture was heated to a high temperature, the bricks were much stronger.

The basic method of making bricks has not fundamentally changed.

Variety

Whether made from clay, sand and lime or concrete, bricks may be divided into four broad varieties: facings, commons, engineering bricks and refractories.

Facing bricks

These are made in a wide variety of colours and surface textures so as to be durable and attractive to look at.

Commons

These are for general-purpose walling that is most likely to be below ground level, externally rendered or internally plastered. They are not given particularly attractive surface features, but are hard and durable.

Common bricks have been largely displaced by lightweight building blocks for internal partition walls and the inner leaves of cavity walling.

Engineering bricks

These are exceptionally hard, dense bricks which have a low porosity and therefore absorb very little water.

Engineering bricks are intended for walls that are heavily loaded, or very exposed to a risk of frost damage. They were originally developed by brick makers in Victorian times, in response to requests by civil engineers for a very strong brick for use in tunnels, bridges and viaducts.

Refractories

These bricks are made from specially selected clays that will withstand very high temperatures.

Quality

There are three qualities of brick:

- Ordinary – Ordinary quality bricks are durable enough to be used in the external face of a building. They can resist frost attack and there is no limit on their soluble salt content.

- Internal – Internal quality bricks need to be protected when used externally. There is no limit on their soluble salt content and they need not be frost resistant.

- Special – Special quality bricks must be durable enough to withstand harsh weather conditions, where they will be constantly soaked with water and attacked by frost. They must have a limited soluble salt content.

Type

There are five main types of brick (Figure 4.49):

- Solid – The volume of pores in a solid brick must not be greater than 25 per cent of the total volume of the brick.

- Frog – A frog is the depression formed in one or both bed faces of a brick. The volume of the frog should not be greater than 20 per cent of the total brick volume.

- Hollow – The volume of the larger holes should be more than 25 per cent of the total brick volume.

- Keyed – Dovetail grooves in one header and one stretcher face provide a key for plaster or rendering

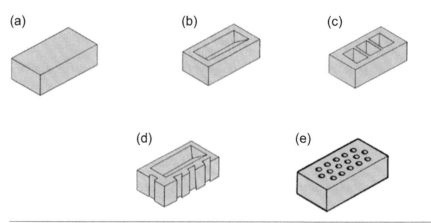

FIGURE 4.49
Types of brick: (a) solid; (b) frog; (c) hollow; (d) keyed; (e) perforated

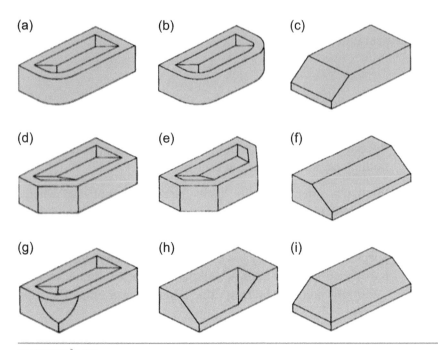

FIGURE 4.50

Types of purpose-made brick: (a) single bullnose; (b) double bullnose; (c) plinth header; (d) cant brick; (e) double cant; (f) plinth stretcher; (g) bullnose stop; (h) internal plinth return; (i) external plinth return

- Perforated – The volume of the small holes should be greater than 25 per cent of the total brick volume.

Calcium silicate bricks

More commonly called sand lime bricks on site, these are made from a carefully controlled mixture of 90 per cent fine silica sand plus 10 per cent lime.

This mixture is pressed into steel moulds and then steam hardened.

Coarse flint sand is also used to make flint lime bricks, which after steam curing have a higher compressive strength.

Concrete bricks

These are cast from a mixture of fine aggregate and Portland cement pressed into steel moulds. The natural setting and hardening process of cement determines the compressive strength of these bricks. They may be solid or frogged and are available as facings, commons and engineering bricks.

Refractories

More commonly referred to as firebricks, these can be considered as two sorts:

- very dense, cream-coloured, solid bricks used to contain the fire in furnace linings, cement kilns, ships' boilers and for lining steel ladles transporting molten metal in a steel works

- very lightweight bricks used to 'back up' dense refractories, or to insulate chimney shafts to prevent heat escaping from flue gases. (These are the only bricks that will float in water.)

A standard size dense firebrick of 230 × 114 × 76 mm is accompanied by a range of arch shapes, wedges and bullnose bricks made to suit circular section kilns and boilers.

Cellular blocks of this same dense refractory material are used to protect the steel deck of the brick-transporter cars used in tunnel kilns.

Purpose-made bricks

Many buildings have decorative features which require shaped or moulded bricks. Manufacturers keep a standard stock of bricks for this purpose, some of which are illustrated in Figure 4.50. If an unusual shape is desired, the manufacturers will make the bricks to the architect's specification.

The advantages of using a purpose-made brick are that labour is saved in cutting, waste is avoided and the natural surface is maintained.

Blocks

A block is described as a walling unit exceeding the dimensions for bricks. Its height should not exceed either its length or six times its width.

Blocks are produced from clay and concrete.

Concrete blocks

Concrete blocks are produced in a range of shapes and sizes.

The face side is usually 450 × 225 mm, the thickness varies from 37 to 225 mm and the weight varies from 6.3 to 15 kg.

They are produced in solid, hollow and multicut format. Multicuts enable a bolster cut to be made without wastage.

Special blocks

Special blocks such as the return block are usually designed to stiffen walls where bonding could cause weakness.

For closing cavities the reveal block could be used, while another special is used to produce a splayed reveal.

Some manufacturers produce blocks with an insulant bonded to the outside face, while others produce hollow blocks with an insulant inserted in the voids.

Walling built with precast blocks may be divided into two main categories:

- load-bearing
- non-load-bearing.

Load-bearing

These blocks are precast in moulds and compacted with the aid of vibration, or moulding machines involving the use of compressed air, or a combination of both.

These blocks are usually made of concrete comprised of Portland cement and a variety of aggregates, such as crushed stone, rock ballast or shingle.

Non-load-bearing

These can also be precast in moulds, or can be produced in slab format and cut to size when set.

These blocks are usually made with cement and a variety of lightweight materials, such as fly ash or burnt coke.

Foundation blocks

These are manufactured in widths from 250 to 335 mm. These blocks are used below ground level and are designed to support cavity walls. They may be dense or lightweight. The dense ones may require two people to bed them.

The current Building Regulations state that hollow blocks must have an aggregate volume of not less than 50 per cent of the total volume of the block calculated from its overall dimensions.

Hollow blocks must have a resistance to crushing of not less than 2.8 N/mm^2, if the blocks are to be used for the construction of a wall of a residential building having one or two storeys.

In all other circumstances blocks shall have a resistance to crushing of not less than 7 N/mm^2.

Clay blocks

Clay blocks are manufactured in a similar way to clay bricks, using the extrusion/wirecut method. These blocks are made from finely washed clay with certain special properties, which is forced through an extruding machine, in the process of which the blocks are cut off to length as the continuous length of clay emerges from the machine.

The thickness of the walls of the blocks is about 12 mm and this allows the blocks to dry quickly and thoroughly. The green clay blocks are then burnt at a high temperature.

The blocks are usually 300 mm long and 225 mm high and the thicknesses range from 37 to 100 mm for partition walls.

They are available with a smooth face and with dovetailed slots to provide a key for the plaster.

Clay blocks are not very easy to cut, so the manufacturers produce special half-units, enabling the bond to be formed without cutting any blocks.

Table 4.1 shows various types of block.

Remember

Never use bricks and blocks in the same wall.

Table 4.1 Types of block

Type of block	Uses	Material	
Lightweight varieties	Internal non-load-bearing walls and partitions generally	Breeze or clinker; waste coke or ash and cement; burnt clay	
Dense heavy	Internal load-bearing walls and external walls	Usually concrete	
Hollow concrete	External walls, usually rendered	Concrete	
Cellular (lightweight)	Load-bearing internal walls	Burnt clay	

Specials

Many block manufacturers provide specials cut to assist in the bonding on site and to prevent wastage from cutting (Figure 4.51).

Stone

Stone is a natural material.

Natural stone and rocks are used in many construction products as walling, roofing, paving and floor tiling products.

Stones can be classified as:

- igneous
- sedimentary
- metamorphic.

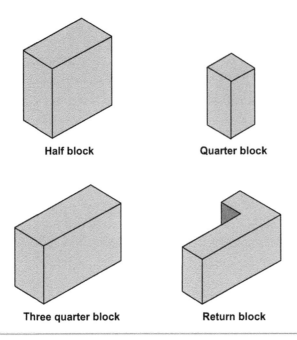

FIGURE 4.51
Special blocks

Igneous stones were formed by the original solidification of the earth's crust. They produce a very dense material, with granite being the main type.

Sedimentary stones were formed by the settling process, when particles were deposited in layers and gradually became compacted to form the stone. Sandstone and limestone are the main types used in construction.

Metamorphic stones were formed from older stones that had been subjected to very high pressures and temperatures. This caused a structural change in the material. Marble, which was formed from limestone, is the main material in this category.

Insulation

Numerous materials can be included into the design of buildings to restrict the passage of heat through the external walls of the property.

Basically building materials are poor insulators of heat. A good conductor of heat is still air and most insulators contain air traps to provide their insulating quality.

Many types of insulating materials are available, each having certain properties.

Fibreglass

Fibreglass consists of fine strands of glass woven together to form sheets or rolls. The sheet form can be used for insulation cavities and the roll form used for insulating roof spaces.

Polystyrene

This is an expanded plastic resin produced in sheet form for use in cavity walls and loose granular fill for cavity walls or roof spaces.

Rock fibre

This woven material is used for both cavity walls and roof spaces.

Vermiculite

This is produced in a flaky form and sold in bags. It is used as a loose fill for roof spaces. It can also be added to concrete to produce a lightweight concrete with improved thermal insulation and fire-resisting properties.

Characteristics of materials

Each of the materials dealt with has its own characteristics, the most important of which are:

- compression
- density
- durability
- flexibility
- strength
- porosity
- conductivity
- expansion.

Compression

When materials are supporting loads they are mainly under compression, i.e. being squashed.

Concrete is very strong under compression, as are bricks.
Insulation materials are very poor under compression.

Density

We are all aware that some materials are heavier than others.

A brick, for instance, is heavier than a slab of polystyrene insulation of the same size. This weight is known as its mass and is measured in kilograms (kg) or grams (g).

Density is the amount of substance in a material. The density of a material is found by dividing its mass by its volume, e.g. kg/m^3.

Durability

This is a general term relating to the ability of a material to resist wear and tear. If a material takes a long time to wear away it is said to be durable.

The durability of certain materials will change depending on their position. For example, plasterboard has good durability inside a building but very poor durability on the exterior of a building.

Bricks and concrete have excellent durability.

Flexibility

Flexible materials can be bent easily without breaking.

Good examples of flexible materials are copper and lead.

A material that is not very flexible is glass in its formed state.

Strength

All materials must be capable of safely supporting their own weight, dead loads and any superimposed loads, such as weather. These loads should not cause any distortion of the structure.

The strength of a material can change according to its physical properties. Some materials become weaker when damp.

When a material such as brick or concrete is being crushed this is known as being compressed or under compression. Conversely, when a material is being stretched it is said to be under tension.

Concrete is strong in compression but weak in tension. If concrete is being designed to be stronger under tension steel reinforcement bars are added, steel being strong in both tension and compression.

Porosity

Many building materials contain water, which either was left there when the material was produced or has been drawn into the material by capillary action.

Timber has a certain amount of moisture in it when it is used, but this occurs naturally.

Concrete and bricks are examples of materials with water left in them from their production.

As this water dries out it leaves pockets of air or voids. Materials with such voids are said to be porous. When they come into contact with water they absorb it by capillarity. Capillarity is the action of the water being sucked into the material along the small voids.

This could cause permanent damage to external materials such as bricks. For example, the face could be blown off.

Conductivity

Buildings are required to be kept at a reasonable temperature so that the occupants feel comfortable.

Heat will flow out of a building whenever the external temperature is less than the internal temperature.

Heat flows through solid materials by thermal conductivity. This is a measure of how good a material is at conducting heat.

The more air voids in a material the better the insulator, as with fibreglass quilt.

Expansion

Materials generally expand when heated and contract when cooled.

Large expanses of brickwork could crack under expansion, so expansion joints are built into the walls to counteract it. Concrete suffers from the same problem and again expansion joints are built in.

The actual amount of expansion varies according to the type of material and the thermal conditions. All materials have a very small amount of expansion.

Plastic guttering provides a perfect example of movement when the weather becomes warmer. You can hear the material making a noise as it moves across the brackets.

If correctly designed a building should not become dangerous due to expansion of materials.

Building defects

If a building causes problems after it has been completed it can be very costly to the client.

Poor design accounts for 7 per cent of the work of the construction industry, working to correct these faults.

Defects in the structure of the building can be caused by:

- poor design
- external effects.

Poor design

Poor design could be attributed to the wrong use of a material, poor supervision of the work or the use of substandard materials.

If materials are used in situations where they are not designed to be used, they will not last as intended. A typical example is common bricks being used in an inspection chamber when engineering-quality bricks should be used.

Poor supervision could lead to the jointing of face brickwork not being carried out correctly and dampness being allowed to penetrate the wall.

Poor storage of facing bricks could result in efflorescence at a later date.

If poorer materials are used instead of the specified quality they will fail to reach their required standard.

External effects

External effects could be attributed to weather conditions, poor supervision and substandard materials being used.

Dampness causes rapid deterioration of materials. When materials become damp inside a building this could lead to other problems such as fungal decay, insect attack, mould growth and efflorescence.

Rain could penetrate a building through cracks, joints and poor application of weatherings.

Lack of control by the site supervisor could also lead to problems with poorly fitted components and mastic seals.

Dampness

It is important that the bricklayer understands the causes of dampness and the methods of damp prevention in the construction of a brick wall.

Dampness may occur even after precautions have been taken, if precautions are based on insufficient knowledge resulting in lack of care in the application of damp prevention.

Many problems with damp are cause by poor workmanship.

Figure 4.52 shows a dirty cavity with mortar droppings on the wall tie, which create a path by which dampness can enter the building.

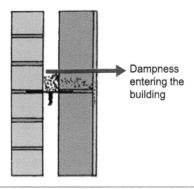

Dampness entering the building

FIGURE 4.52
Mortar droppings on the wall tie causing a bridge

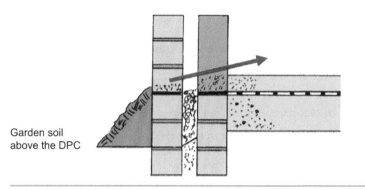

FIGURE 4.53
Garden soil above the DPC level and mortar dropping in the cavity

Figure 4.53 shows garden soil above the DPC level and mortar droppings in the cavity. These form a bridge for dampness to enter the building.

Moisture

Building materials that are porous will deteriorate when wet.

This can cause defects such as swelling, warping and twisting, which cause a reduction in strength and could eventually lead to collapse.

All building materials should be correctly stored before use and protected until the building is watertight.

Chemical attack

When building materials are in contact with the atmosphere they may become corroded.

The atmosphere contains fine particles of dust that pollute the air, especially near industrial areas. These become diluted when mixed with rainwater and can attack the surfaces of unprotected building materials.

The corrosion of metals through rusting and the spalling of stone and brick are the main causes of chemical attack.

Fungal attack

The most common fungal attacks occur on timber products.

There are various forms of fungus which, given suitable conditions, will attack timber until it is destroyed. These can broadly be classified into two groups:

- wet rot
- dry rot.

Wet rot

Wet rot is dark in colour and needs a source of moisture, otherwise it will die.

The affected timber is white or brown in colour and is cracked, although much less so than timber affected by dry rot. The timber becomes brittle, loses its strength and colour, and crumbles easily.

Wet rot is found in places that are constantly damp, e.g. wet timbers in leaky roofs, damp cellars, window frames, around water tanks and under ground-level floors.

Timber affected by wet rot should be cut out and replaced and the surrounding timber treated.

Dry rot

This is the most common and serious fungus. Dry rot is off-white in colour, fluffy like cotton wool, and forms a continuous sheet over the timber. Once formed, dry rot produces its own moisture.

Timber that has been attacked by dry rot darkens in colour and forms cracks along and against the grain.

The timber becomes less dense, becomes brittle and loses its strength. It crumbles easily into a powder.

Dry rot is often found in unventilated places where humidity is greater than 20 per cent; for example, in cellars and cupboards, under floors and behind panelling.

Timber affected by dry rot must be replaced and the surrounding timber treated.

Insect attack

Most insect attacks on timber are by beetles.

A typical life cycle of a beetle occurs in four stages:

- Egg – The female beetle lays her eggs on the surface of bare timber, usually in crevices or cracks.
- Larva – The eggs hatch into larvae (grubs), which bore into and feed on the timber, leaving unwanted bore dust behind.
- Pupa – The larvae change into pupae near the surface of the timber and, during this dormant stage, develop into adult beetles.
- Adult beetle – The adult beetles emerge from the timber through a small hole known as the flight hole and fly away to mate.

Common furniture beetle

The common furniture beetle is responsible for most damage caused in the UK. Commonly known as woodworm, this beetle attacks most softwoods and the sapwood of hardwoods.

It is most often found in furniture and joinery. The eggs are oval in shape and the flight holes circular with a diameter of 1.5–2 mm.

Powder post beetle

The powder post beetle attacks hardwoods and is found particularly in timber yards and sawmills.

The eggs are laid in the large pores of hardwoods and the circular flight holes are about 2 mm in diameter.

House longhorn beetle

This beetle attacks the sapwood of softwoods only and is mainly found in structural timbers. Extensive damage can be caused without its being noticed because of the long larval stage and very few flight holes. The flight holes are oval and 5–10 mm in diameter.

Death-watch beetle

The death-watch beetle attacks hardwoods, usually timbers of old buildings where decay has started. The flight holes are circular and about 3 mm in diameter.

Protection

Wood preservation and protection

Except where timber and wood products are used in their natural state, for environmental, practical or economical reasons, they are generally treated with:

- paint
- water-repellent stain
- wood preservatives
- special solutions to reduce or retard the decay process.

Material records

It is generally accepted that the initial selection of materials plays a major part in achieving a satisfactory building. Unfortunately, it is less recognized that the methods adopted on site for the handling and storing these materials are equally important.

Manufacturers and suppliers go to great lengths to provide materials and components that comply with recognized standards, but much of their effort can be eroded if the same importance is not attached to the product when it is transferred to the care of the construction team.

The cost of materials used in the construction industry exceeds 50 per cent of the overall costs. The supervisor has little or even no control over the initial price of materials and every contractor usually has to pay approximately the same price. However, by employing strict methods of control and checking, efficient handling and correct use of the materials, the most efficient contractor will expend the least cost in their use.

Principles of materials handling and control

Materials handling and control can be divided into the following sectors:

- planning material deliveries
- ordering the materials
- checking deliveries on site
- storing materials on site
- processing delivery tickets and invoices
- controlling the use of materials.

Planning material deliveries

Planning of deliveries usually takes place in the form of charts, graphs, data sheets, diagrams or the written word. It is essential to have a record of the proposed delivery so that reference can be made to it in the future.

However, it must be stressed that these only give a guide and prompt the site manager. Adjustments should be made according to site requirements.

Ordering the materials

The majority of orders for the supply of building materials and components are placed by the main contractor. The main contractor is solely responsible for correct placing of material and component orders, ensuring delivery and making payment for said goods.

Before any order is placed a quotation should be obtained. In a competitive market it is wise to obtain a number of quotations for the supply of a particular item. In this way variations to the price range, delivery dates, conditions and general service can be compared between suppliers, and after careful consideration a quotation may be accepted.

When a supplier is asked to provide a quotation for a particular material or component the main contractor must provide the supplier with as much information as possible to determine the exact specification of the items.

This information can be given by any or all of the following:

- extracts from the bill of quantities
- extracts from the specification
- reference to British Standards
- detailed sketch drawings
- reference to manufacturers' literature or catalogues.

In order to provide a true price the supplier must know when the material is required. When a comprehensive picture has been built up the supplier can prepare a quotation based on reality and not an assumption.

MATERIALS ORDER		WEST BUILDERS LTD		
Date:				
Deliver:		Contract		
Reference Number	Description of materials	Quantity	Rate	Price

FIGURE 4.54
Order sheet

DELIVERY NOTE		ABC BUILDING MATERIALS SUPPLIERS
Your Ref:		
Our Ref:		
Order No:		Date: ....
Invoice address:		Delivery address : . ...
Description of goods	Quantity	Catalogue No
Comments:		
Date and time of receiving the goods:		
Name of recipient: .... Signature:		

FIGURE 4.55
Delivery note

Contractor's orders

When the quotation has been accepted the main contractor will place an order with the supplier similar to the one shown in Figure 4.54. This should include the company heading, contract name and delivery address. It should also give a full description of the materials as described in the contract documents.

Checking deliveries on site

When the materials are delivered to the site they should be accompanied by a delivery ticket. This ticket should be checked against the quantity and type of material. You should also check for any damage and inform your supervisor immediately. A typical delivery ticket is shown in Figure 4.55.

A material schedule sheet will have been completed before the site was set up, and gives the site manager the information required for controlling the materials on site.

If materials arrive on site very early they are more prone to damage or pilfering.

Storing materials on site

The storage of materials must be considered during early planning.

Depending on the availability of space and the size of the contract, properly fitted-out stores, with racks and bins of varying sizes into which materials can be placed for ready issue and easy handling, should be available under the charge of a responsible, competent storeperson.

The stores should lie within an enclosed stores compound. The compound should have adequate space and provision on which to place materials that cannot be accommodated within the stores and are not affected by the weather. These stores must be kept to a minimum for security reasons.

It is important that stores are strictly controlled and only authorized employees allowed to sign for receipt. The stores should have a receiving bay and entrance door, and an entirely separate entrance for the issuing of stores, complete with a counter and register for outgoing materials. This register should indicate for which section or part of work the materials are required, and under whose authority they are issued.

Tarpaulins, plastic sheeting or protective coverings should always be available for issue.

Processing delivery tickets and invoices

The site manager should receive some notification of impending deliveries. This could be an advice note from when the order was originally made.

The materials should be accompanied by a delivery note, presented by the driver as proof of delivery, quality and quantity. Meticulous checks should be made of the materials delivered for quality and quantity. Delivery notes should not be signed until the recipient is satisfied of the quality and quantity of the goods.

```
┌─────────────────────────────────────────────────────────────┐
│                    WEST BUILDERS LTD                         │
│                    MATERIALS RECEIVED                        │
│                                                             │
│  Contract: ...........................................       │
│  Job No:    .......................................          │
│  Prepared by: .......................... No:............     │
```

Date	Delivery No	Supplier	Materials	Rate	Total Value	Comments

FIGURE 4.56
Materials received sheet

```
┌─────────────────────────────────────────────────────────────┐
│              ABC BUILDING MATERIALS SUPPLIERS                │
│                                                             │
│                         INVOICE                             │
│                                                             │
│   No:  ...............................                       │
│                                                             │
│   Messrs: ...............................                    │
```

Invoice date	Your order No	Dated	Sent	Code
....	....	....	....	....

Quantity	Description	Price	Items	Total

No claim can be entertained unless made three days after receipt of goods.

FIGURE 4.57
Typical invoice

Each individual delivery ticket should be sent to the head office and a record of the delivery kept on site on the sheet shown in Figure 4.56.

Once the material has been correctly delivered the supplier will send an invoice to the building company requesting payment. A typical invoice is shown in Figure 4.57.

WEST BUILDERS LTD MATERIALS RECEIVED		
No: .... **Date:**		
From site:	**To site:**	
Description	**Quantity**	
Issued by:	**Received by:**	**Driver's signature**

FIGURE 4.58
Materials transfer sheet

Material transfers

When materials need to be transferred between sites, correct documentation should be completed (Figure 4.58).

Controlling the use of materials

Careless off-loading and handling of materials, goods and components adds to wastage on the construction site. Each site manager should pay adequate attention to the methods used depending on the types of material being dealt with.

Secure accommodation together with an effective storage system is essential on construction sites to prevent unnecessary waste and loss of materials due to vandalism and theft, and damage caused by constant handling.

Storage control

Irrespective of the amount of materials being stored, the following factors need to be considered.

Deliveries

Materials received need to be checked against suppliers' delivery advice notes for correct quantities and damage.

Stock rotation

Stocks of some materials (plaster, cement, paint, etc.) need to be used in the order of delivery as they have only a limited shelf-life (Figure 4.59).

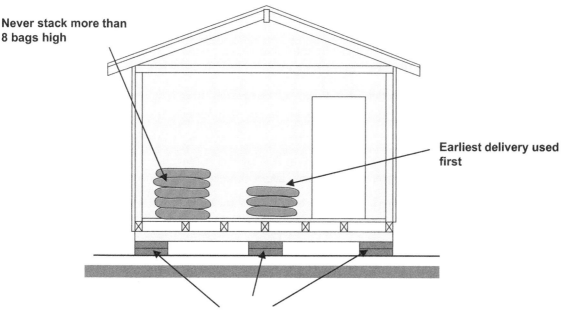

Never stack more than 8 bags high

Earliest delivery used first

Waterproof shed raised off the ground on battens on blocks

FIGURE 4.59
Stock rotation of bagged materials

Withdrawals

Materials drawn from stock by requisition need to be replaced.

Security

Storage facilities must be capable of being made secure against vandalism and theft.

Cleanliness

The store and unloading areas must be kept clean and free of debris at all times.

Safety

Materials must be handled and stacked in such a manner that injuries and damage are avoided. Hazardous materials need to be stored in separate accommodation constructed of non-flammable materials.

Minimizing waste

It is recognized that wastage of materials will occur on a construction site for many reasons; therefore estimators, when pricing for work, allow within their rates a wastage percentage ranging from as little as 1 per cent to an alarming 15 per cent.

On different construction sites, the waste of materials for similar operations varies considerably. This difference is without doubt caused by differences in the control exercised by the site supervisor.

Wastage of expensive materials, in particular those in short supply, can make the difference between profit and loss.

Site management must, therefore, pay particular attention to the prevention of waste by instituting regular inspections and careful assessments of material requirements.

With timber and reinforcement the preparation of cutting lists before actual ordering takes place, combined with correct storage, can avoid unnecessary wastage.

With the growth in the industry of labour-only subcontractors, wastage should be particularly controlled. If, in the early stages of the work, it is found that wastage is taking place owing to carelessness on the part of the subcontractors, a quick reminder of a 'contract charge' is most effective.

Strict stores control can often pinpoint at an early stage who is responsible for calling up materials and then leaving them around the site.

Multiple-choice questions

Self-assessment

This section of the book is designed to allow you to check your level of knowledge. The section consists of revision questions for this chapter. The questions are all multiple choice and have four possible answers. The answers are to be found at the end of the book.

The main type of multiple-choice question will be the four-option multiple-choice question. This will consist of a question or statement, known as the stem, followed by a choice of four different answers, called the responses. Only one of these responses is the correct answer; the others are incorrect and are known as distracters.

You should attempt to answer the questions by choosing either (a), (b), (c) or (d).

Example

The person employed by the local authority to ensure that the Building Regulations are observed is called the:

    (a) clerk of works

    (b) building control officer

    (c) council inspector

    (d) safety officer

The correct answer is the building control officer, and therefore (b) would be the correct response.

Construction technology

Question 1 Which of the following should you never build in front of?

    (a) the frontage line

    (b) the brick line

    (c) the setting-out line

    (d) the building line

Question 2 Which of the following machines is best suited for trench excavations?

    (a) backactor

    (b) front shovel

    (c) drott

    (d) scraper

Question 3 The type of foundation shown is known as a:

(a) raft foundation
(b) strip foundation
(c) pad foundation
(d) pile foundation

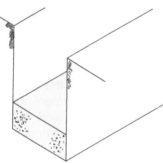

Question 4 Identify the type of insulation shown:

(a) expanded polystyrene
(b) fibreglass
(c) polyurethane foam
(d) vermiculite

Question 5 The type of roof form shown is known as a:

(a) flat roof
(b) lean-to roof
(c) gabled roof
(d) hipped roof

Question 6 Identify the type of brick shown:

(a) engineering brick
(b) frogged brick
(c) solid brick
(d) perforated brick

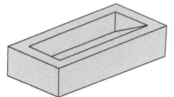

Question 7 Identify in what type of soil would the timbering shown be used:

(a) soft soil
(b) loose soil
(c) moderately firm soil
(d) firm soil

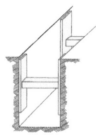

Question 8 The most common fungal attack on timbers is known as:

(a) woodworm
(b) efflorescence
(c) dry rot
(d) mould growth

CHAPTER 5

Moving and Handling Resources

This chapter will cover the following City and Guilds units:
- Apprenticeship   Module 9   S4.1, S4.2, S4.3, S4.4
                    Module 12   S1.3
- Certificate   Unit 201   3.1, 3.2, 3.3
- Diploma   Unit 201/601   5.1, 5.2, 5.3

This chapter is about:
- Interpreting building information
- Adopting safe and healthy working practices
- Selecting materials and components according to characteristics and limitations of use
- Handling, moving and storage of materials and components by manual and mechanical methods

The following Skills and Knowledge requirements will be covered:
- Moving materials using lifting techniques

The following Certificate learning outcomes will be covered:
- Understand how to set up and secure work areas
- Consideration for areas of welfare

The following Diploma learning outcomes will be covered:
- Know how to handle materials and equipment safely
- Understand lifting aids

Complying with product information

Material characteristics

Construction materials can be classified into the following groups:

- bulk

- hazardous

- fragile or perishable

- miscellaneous.

Bulk material

Bulk means a large amount of one type of material, e.g. bricks, sand and aggregates.

They are delivered to site in large amounts to save on delivery charges. Some material suppliers deliver loose bulk material in bags.

The following are examples of bulk materials (Figure 5.1):

- bricks, blocks, roof tiles, concrete units, drainage pipes

- aggregates – sand, shingle and granolithic chippings

- timber – hardwoods and softwoods.

Hazardous materials

These materials usually have instructions written on their containers or packages explaining the handling and storing requirements.

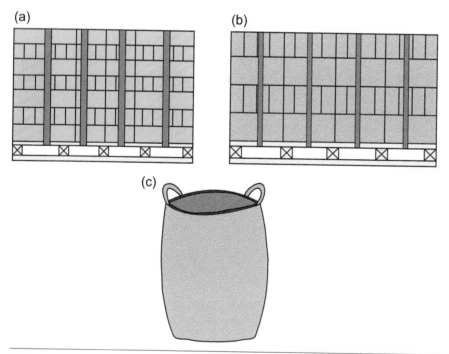

FIGURE 5.1

Various bulk materials: (a) pallet of bricks; (b) pallet of blocks; (c) bagged material

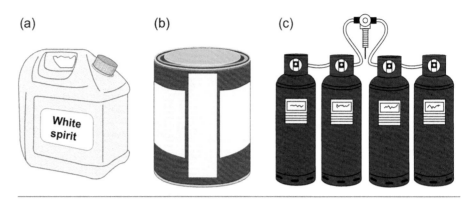

FIGURE 5.2
Various hazardous materials: (a) white spirit; (b) paint; (c) gas bottles

Always ensure you read the instructions, and wear the appropriate personal protective equipment (PPE) before handling hazardous materials.

The following are examples of hazardous materials (Figure 5.2):

- paints and solvents
- mastics
- cleaning agents
- industrial gases – butane and oxyacetylene
- highly flammable liquids.

Fragile or perishable materials

Many of the materials and components delivered to building sites are durable and can be stored out in the open; for example, sand, aggregates, bricks and blocks.

Other items are perishable and fragile and require special attention when handling and storing.

Durable means that a type of material or component is hard wearing and can withstand extreme weather conditions when used for its intended purpose.

Perishable items such as cement require special storage procedures to ensure the shelf-life of the material.

The following are examples of fragile or perishable materials (Figure 5.3):

- sheet materials – plywood, blockboard, chipboard, plasterboard
- glass – vitreous china – WC pans, basins, etc.
- glazed tiles
- roofing felts – damp-proof course (DPC) materials, pitched fibre and polythene products.

Several of these materials need to be unloaded, transported and stored under dry, secure conditions on building sites to prevent deterioration and loss.

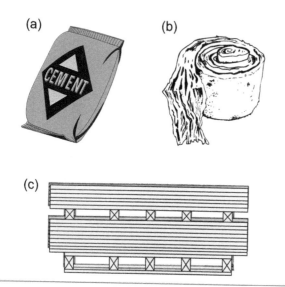

FIGURE 5.3
Fragile or perishable materials: (a) cement; (b) fibreglass; (c) plasterboard

Miscellaneous

The following are examples of miscellaneous materials (Figure 5.4):

- ironmongery – nails, screws, hinges, door furniture
- doors – hardwood and softwood
- units – kitchen and bathroom
- windows – wood, metal and PVC.

All the above materials need to be stored under dry, secure conditions on building sites to prevent deterioration and loss.

This will be dealt with later in the chapter.

Material identification and limitations of use

Aggregates

Aggregates are particles or granules of materials that are used with a binder and water in controlled proportions to produce a solid mass when set. By the use of differing aggregates and binders, various types of materials can be produced.

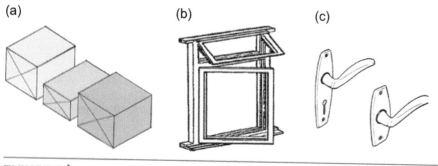

FIGURE 5.4
Miscellaneous materials: (a) packages; (b) window frame; (c) ironmongery

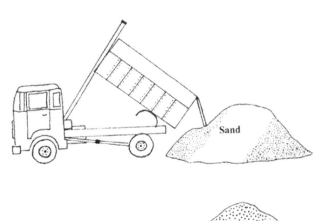

FIGURE 5.5
Bulk materials arriving on site by lorry

Aggregates can be divided into two further categories.

- fine aggregate or sand – used for making cement mortar

- coarse aggregate – used in the production of concrete.

Fine aggregate or sand has very small particles. It originates from either quarries or river beds.

Aggregate may arrive on site in lorries for large quantities or bagged for small quantities (Figure 5.5). It is also available in various colours.

Coarse aggregate is crushed rock or pebbles which originate from quarrying stone and are then crushed to achieve the required size of aggregate.

- Sand

    - soft sand – used in mixing mortar for bricklaying

    - sharp sand – mixed with ballast/cement to form concrete.

- Shingle

    - pea-gravel – used as a fine granular filling material, e.g. around drain pipes

    - premixed sand/shingle – sharp sand/shingle used for floor screeding and fine concreting

    - ballast – larger aggregates used with sand for mass concreting, i.e. foundations.

- Granolithic chippings – used where hard-wearing textured surface is required, e.g. garage floors, door thresholds.

- Hardcore – usually graded limestone, but broken or crushed clean brick or concrete could also be used.

Bagged materials

Bagged materials arrive on site in specially prepared bags. As well as taking care not to damage them when unloading, the storage is of special importance owing to their excessive deterioration through dampness.

Portland cement

Produced in various grades, including: ordinary, rapid hardening, super sulphate and masonry cement. They arrive on site in 25 kg bags. The bags are produced from several layers of light brown paper, one of which has been treated to prevent moisture reaching the dry cement powder.

Lime

Manufactured from limestone and used for bricklaying and plastering.

Lime arrives on site in the same way as cement and is also in bags, but white in colour.

Pallets are used on site to prevent rising damp attacking the dry powder of the lime and cement and causing it to go hard and unusable (Figure 5.6).

Plasters also arrive on site in bags and should be stored in the same way.

Timber materials

Timber – sheeting materials such as plywood and blockboards could be delivered either in packs or individually. Carcassing timber could also arrive either in packs or individually (Figure 5.7).

Timber is a natural material.

Timber is used for many purposes in which its strength is a matter of critical importance.

Historically, timber has been classified into:

- hardwoods
- softwoods.

There is little correlation between these terms and the actual hardness and softness of the timber.

Softwoods

Softwoods are produced from coniferous trees, which are evergreens and grow chiefly in the northern temperate zones. They have cone-like leaves.

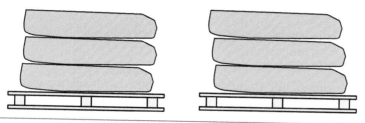

FIGURE 5.6
Bagged material on pallets

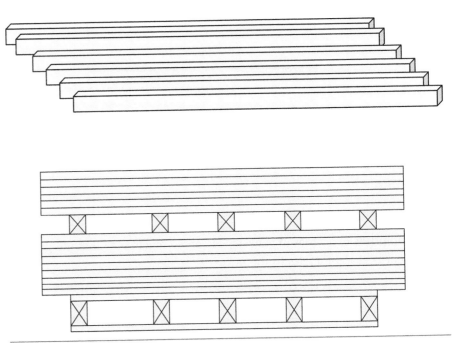

FIGURE 5.7
Timber delivered loose and in packs

Hardwoods

Hardwoods are produced from deciduous trees with broad leaves, although holly, certain oaks and the majority of tropical trees are evergreens.

Hardwoods and softwoods are botanical terms and do not always relate to hardness. Thus, balsa is a hardwood, while yew, an extremely hard timber, is a softwood.

Sheet materials

Boards

In its natural form timber is one of the most useful and versatile of materials used on the construction site. It has, however, certain inherent shortcomings that tend to limit its usefulness in its solid form, i.e. as planks, boards, etc.

These shortcomings include the following:

- Width of boards – In its natural form a board is limited to the log from which it is cut.

- Stability – Timber, especially in large sizes, is likely to warp, shrink or swell.

- Strength – Timber is weak across its grain.

Artificial boards attempt to overcome these shortfalls.

Plywood

Plywood is available in large, standard size sheets in a range of thicknesses from 1 mm upwards (Figure 5.8).

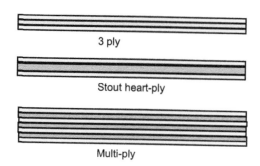

FIGURE 5.8
Types of plywood

It comprises of a series of veneers, cemented together so that the grain of one veneer is at right angles to the next.

Particleboards

Generally referred to as chipboards, these are made of wood chips and flax chives, which are bonded together with a synthetic resin adhesive.

The boards are available in lengths of up to 5.3 m, widths of up to 1.7 m and thicknesses from 8 to 40 mm. There are also various densities and grades available. An example is shown in Figure 5.9.

Hardboard

This is available in standard size sheets from 3 to 6 mm in thickness and is used for the same purpose as plywood, being somewhat cheaper, heavier and inferior.

Insulation board

This low-density board is made from the same materials as hardboard, but is not pressed in its manufacture.

Insulation board or softboard is available in various textures and finishes to suit the wide variety of work for which it can be used.

Thicknesses range from 6 to 20 mm and it can be used in either sheet or tile form.

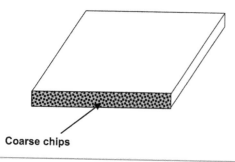

Coarse chips

FIGURE 5.9
Example of particle board

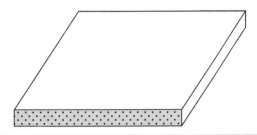

FIGURE 5.10
Example of medium-density fibreboard (MDF)

Medium-density fibreboard (MDF)

Fibreboard is basically a compromise between hardboard and softboard (Figure 5.10).

It has better heat insulation qualities than hardboard while retaining greater mechanical strength than softboard, and is very easy to work.

Available in standard sized sheets, it is normally 9.5 mm in thickness.

Blockboard

This is a very useful sheet material made by sandwiching a core of narrow strips of wood between two face veneers (Figure 5.11).

The best grades of blockboard have double face veneers so that the core strips and the grain of the face veneers run in the same direction.

Blockboard is available in the usual standard size sheets in thicknesses from 12 to 48 mm, and has core strips up to 25 mm wide.

Laminboard

Laminboard is essentially similar in construction to blockboard, but is generally superior and more stable owing to the narrowness of the core strips, which have a maximum width of 7 mm.

Laminboard is also glued edge to edge, unlike some of the cheaper grades (Figure 5.12).

The boards are available from 12 to 25 mm in thickness.

Battenboard

This material is a coarser version of blockboard and has core strips up to 75 mm wide (Figure 5.13).

It is less stable than blockboard and it is best suited for jobs where overall strength and tendency to 'ripple' are less important.

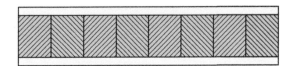

FIGURE 5.11
Example of blockboard

FIGURE 5.12

Example of laminboard

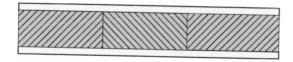

FIGURE 5.13

Example of battenboard

Plasterboard

Plasterboard consists of an aerated gypsum core encased in specially prepared paper liners.

The boards are available in plain or insulating grades, the latter having a veneer or polished aluminium on one side (Figure 5.14).

There are four types of basic plasterboard:

- gypsum wallboard
- gypsum lath
- gypsum baseboard
- gypsum plank.

Plasterboards are available in many sizes, the most common being:

- 1.210 m × 0.900 m × 9.5 mm thick
- 1.829 m × 0.900 m × 12.5 mm thick.

Steel reinforcement

Steel reinforcement arrives on the building site in large sheets or single bars.

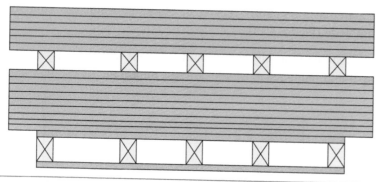

FIGURE 5.14

Example of plasterboard

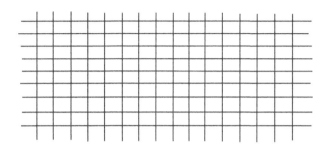

FIGURE 5.15
Example of sheet reinforcement

The sheets consist of small-diameter bars welded together to form a mesh, and are used mainly for reinforcing the foundations and floors of buildings. Sheet reinforcement is coded according to the size of the bars and the size of the mesh (Figure 5.15).

Steel reinforcing bars are available in various diameters according to the strength required, ranging from 6 to 40 mm.

There are also numerous deformed bars such as those shown in Figure 5.16. These improve the bond between the concrete and the steel reinforcement.

There will be occasions when the steel reinforcement is delivered to the site already bent and formed, ready for fixing in the appropriate foundation, column or floor, etc.

In this case it is important that you can recognize the various items of re-inforcement that have been fabricated off site and are able to store them correctly.

Containerized materials

Paint

Paint is delivered to the normal construction site in small quantities, as and when the decorators require it, because it has a very short shelf-life.

Paint and other liquid items should be stored inside a special store and protected. Examples are shown in Figure 5.17.

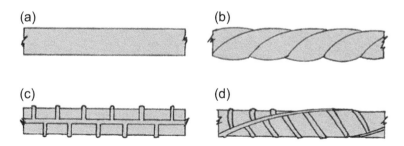

FIGURE 5.16
Examples of bar reinforcement: (a) plain round bar; (b) square twisted bar; (c) ribbed bar; (d) twisted ribbed bar

FIGURE 5.17
Examples of containerized materials

There are many types of decorating materials such as emulsions, undercoat and gloss, as well as varnishes and protective coverings for timber, masonry and steel.

They should all be easily recognizable at the time of delivery.

Petrol, diesel and oil

These items are classified as highly flammable liquids (HFLs) and should be received on site with the utmost care and attention.

They should all be kept separate and never mixed.

The canisters, barrels, etc., should be clearly marked and kept in a protected store (Figure 5.18).

The major hazard associated with these liquids is *fire* and it is essential that precautions are taken to limit the risks involved.

Rolled material

Examples of rolled materials are (Figure 5.19):

- DPC
- polythene (membrane sheeting)
- roofing felt

FIGURE 5.18
Barrels of flammable liquid

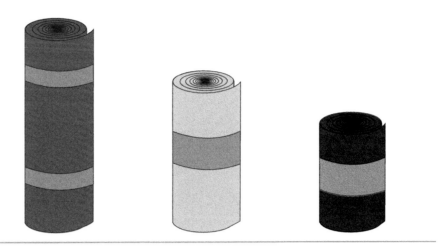

FIGURE 5.19
Examples of rolled materials

All rolled materials should be stored in an appropriate store, on end to retain their shape. They should be stored in their appropriate sizes for ease of identification.

Boxed materials

Materials arriving in boxes include:

- flat packs (kitchen units)
- nails
- ironmongery
- fittings.

All of the above are delivered to site when required as they are usually some of the most valuable materials and are easily stolen.

Budget-priced units are often delivered in knock-down form, known as flat packs, ready for on-site assembly. Flat packs relate mainly to kitchen units, which consist of a range of units mass produced to standard designs by manufacturers. They should be delivered and stored in the unit where they will be fixed.

These items are easily damaged and special care should be taken when unloading and storing them. Figure 5.20 shows an assembly drawing for erecting flat-packed units.

Fixings

Nails, screws, etc., are delivered during the contract and are usually bought in bulk for cheapness.

There are numerous types of screw available and they are ordered according to length and gauge (thickness). There are also various types of screw heads.

Many types of nail are available; these are also ordered according to length and gauge.

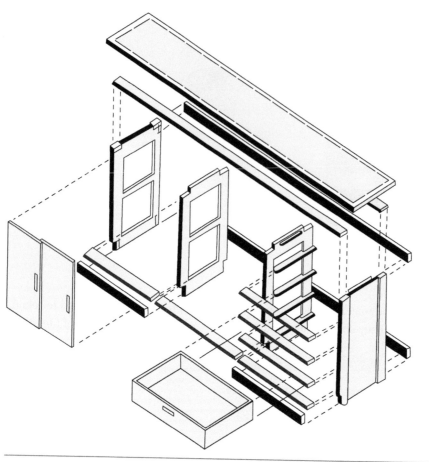

FIGURE 5.20
Flat-packed units

Ironmongery

Carpenter's locks, bolts, handles, etc., are desirable items that are likely to be stolen unless stored securely under the control of a storeperson.

They are supplied as individual items, plastic bubble packed on cards or in boxed sets, e.g. 10 pairs of level handles.

Examples of typical packages are shown in Figure 5.21.

Sources of information

Information on materials can be obtained from numerous sources:

- books and technical articles
- manufacturers' literature
- specifications
- Building Regulations
- British Standards Institution
- British Standard specifications

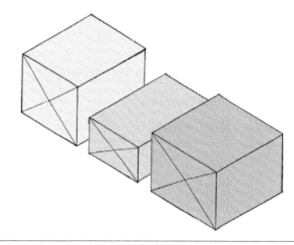

FIGURE 5.21
Examples of packaged materials

- codes of practice
- Building Research Establishment (BRE)
- BRE digests, information papers and current papers.

Manufacturers and trade development associations:

- Brick Development Association
- Timber Research and Development Association
- Fibre Building Board Development Organisation
- Cement and Concrete Association
- Lead Development Association

and many more.

Copies of their publications are normally available for students' reference in most college and town libraries.

Many libraries also take trade periodicals which operate a reader's enquiry service whereby technical brochures and information can be obtained from various manufacturers and suppliers.

Trade periodicals:

- *Cement and Concrete Monthly*
- *The Architects' Journal*
- *Building*
- *The Brick Bulletin*

and many more.

Professional institutions and technical societies are also a source of specific information:

- Chartered Institute of Building

- Royal Institution of Chartered Surveyors

- Institution of Civil Engineers

- Institute of Clerks of Works

and many more.

Interpreting technical information

It is important that the student understands where information can be found. It can be gathered from numerous sources.

Manufacturers' technical information

Any efficient office should have up-to-date information regarding new and existing products from the various manufacturers who produce materials of interest to them.

Many manufacturers produce technical information which is free for the asking. These papers, brochures, leaflets, etc., should be stored in the office in an easily retrievable system.

The easiest way is to file them under each specific trade, or material.

The most elaborate system is called the SfB Classification System. To assist in the filing of literature each item is coded with appropriate SfB references in the top righthand corner of the document (Figure 5.22).

Information can be written, oral or in drawing format.

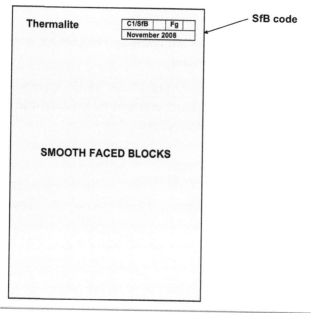

FIGURE 5.22

Typical supplier's information

Oral information

This is received from people such as supervisors and workmates by word of mouth.

Care should be taken to ensure that the person receiving the message has received the correct version of the message, although there will be no real proof of this.

Noise on building sites can cause problems when oral information is being passed from person to person.

However, a quick response can be achieved with oral instructions.

Written information

This is usually preferred to oral communication as the person receiving the message receives the correct version and has documented proof.

A permanent record can be very useful at a later date if problems occur.

However, it takes time for written information to reach to the recipient.

Tables, graphs and charts

Most construction sites have programmes to explain how a contract is progressing. Bar charts are the most common method used.

Charts can also be produced to show the financial status of the contract.

Specification

This is a document giving a written description of materials to be used, and construction methods to be employed in the construction of the building.

Schedule

Information is presented in a tabulated form for items such as:

- doors
- windows
- ironmongery
- decorative finishes
- inspection chambers (manholes).

Textbooks

There are simply thousands of textbooks available on practically every aspect of the construction industry.

They not only explain the various methods of construction but also give typical details and the reasons why particular building methods are used.

Building regulations

These regulations are concerned with safeguarding the safety and health of occupants of completed buildings.

They deal with matters such as safe methods of design, standards of construction, and selection and use of materials.

British standard specifications

These are issued by the British Standards Institution to lay down minimum standards for materials and components (e.g. walling blocks and doors) used in construction and other industries.

British standard codes of practice

Codes of good practice are issued by the British Standards Institution to cover workmanship in specific areas, e.g. building drainage, and brick and block masonry.

Agrüment certificate

These certificates are granted by an independent testing organization, called the British Board of Agrément, stating that the manufacturer's products have satisfactorily passed agreed tests. Subsequent to the granting of the certificate strict quality control has to be continued.

Drawings

These are produced to assist the builder to produce the building exactly as the architect has designed it. Drawings have been dealt with in Chapter 3.

Refer back to Chapter 3 to reinforce your knowledge of the following types:

- sketches – quick and easy to draw and understand

- location drawings – scale drawings used to show the complete details of the building.

Component characteristics and limitations of use

Identifying and selecting components

Bricks

See Table 5.1 and Figure 5.23.

Table 5.1 Types of brick

Type	Composition and properties	Uses
Clay	Composed mainly of silica and alumina and small quantities of lime, iron and manganese	General use
Calcium silica	Made up from sand and lime or crushed flint and lime moulded under pressure and hardened by exposure to steam	On walls to reflect light
Fire brick	Produced from refractory clay having a high fusing point	Furnace work
Modular brick	Metric modular bricks available in differing sizes: 300 × 100 × 100; 200 × 100 × 100 mm	
Common brick	General bricks, which vary in quality and do not have a decorative face	Work below ground level
Facing brick	Produced with a decorative finish applied to the header and stretcher courses	Decorative work
Engineering brick	Hard and dense with a smooth texture, having a high load-bearing capacity	DPCs or supporting heavy loads
Insulation brick	Made from diatomaceous silica to give heat insulation properties	Lining chimneys
Specials	Purpose made to meet some detail of construction e.g. squint, plinth	Features

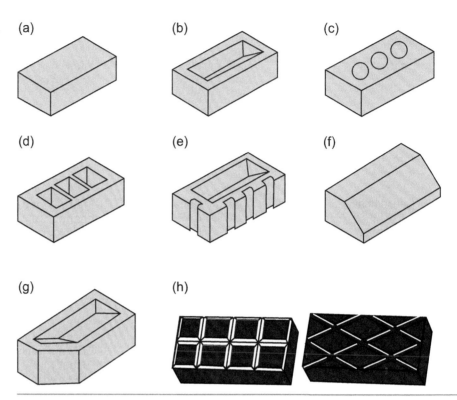

FIGURE 5.23
Types of brick: (a) solid wire-cut brick; (b) pressed or hand-moulded brick; (c) perforated wire-cut brick; (d) cellular pressed brick; (e) keyed pressed brick; (f) plinth stretcher; (g) cant brick; (h) paving bricks

Bricks are usually identified according to:

- place of origin – Nottingham red
- colour – buff, dapple grey
- method of manufacture – pressed, wire-cut, hand-made
- surface texture – sand faced, smooth
- use – engineering brick, facing brick
- size – metric, imperial.

Blocks

Blocks are produced in solid, hollow or multicut form in a range of shapes and sizes (Table 5.2).

Drainage pipes

Drainage pipes can be manufactured from various materials and can be classed as rigid or flexible. A sample of pipes is shown in Figure 5.24.

Clay is the major material used for rigid drain pipes in domestic work, with cast iron as the main alternative. The usual materials for flexible drain pipes are pitch fibre and unplasticized PVC.

Table 5.2   Types of block

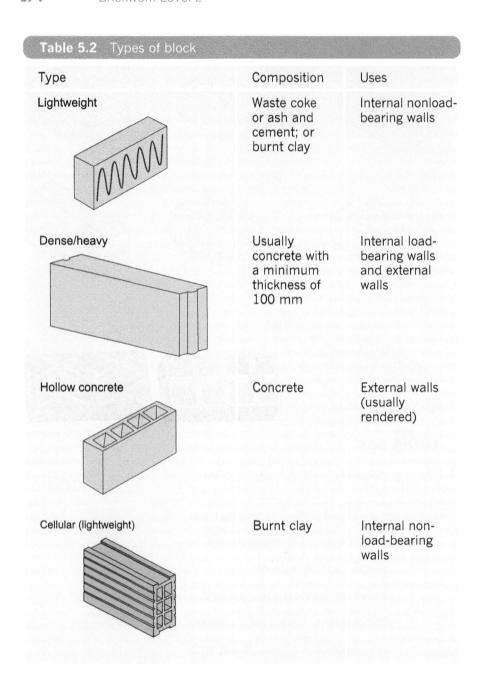

Type	Composition	Uses
Lightweight	Waste coke or ash and cement; or burnt clay	Internal nonload-bearing walls
Dense/heavy	Usually concrete with a minimum thickness of 100 mm	Internal load-bearing walls and external walls
Hollow concrete	Concrete	External walls (usually rendered)
Cellular (lightweight)	Burnt clay	Internal non-load-bearing walls

Concrete units

Concrete is used for the manufacture of numerous items used in construction. Concrete kerbs, paving slabs and lintels are some the main products.

Concrete kerbs

Kerbs are used at the junction between the footway and the carriageway.

A kerb is a specially shaped piece of precast concrete (Figure 5.25). It serves to define the edge of the carriageway and to direct the flow of water from the pavement surface.

The kerb is bedded in concrete to give it stability.

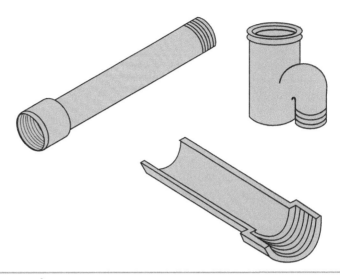

FIGURE 5.24
Types of drain pipe

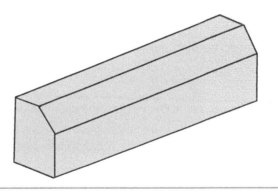

FIGURE 5.25
Concrete kerb

Concrete paving slabs

Concrete paving slabs are used on the surface of footpaths and causeways, but are used more for private footpaths than council causeways.

They are available in various shapes, sizes and colours. Typical shapes are shown in Figure 5.26.

Concrete lintels

Although other materials are more common nowadays the concrete lintel is still used on many construction sites.

FIGURE 5.26
Concrete paving slabs

FIGURE 5.27
Concrete lintel

Lintels are used over openings to transmit loads to the jambs at the sides of the openings. A typical concrete lintel is shown in Figure 5.27.

Roofing tiles

These are coverings for pitched roofs and can be manufactured from either clay or concrete. Various types are shown in Figure 5.28.

They are nailed to timber battens which are fixed to the rafters of the roof. The tiles are hooked over the battens by means of nibs which provide protection from the underside.

The coverings may be either double or single lap.

Flat packs

Many internal fittings are delivered in flat packs, that is they are delivered unfitted, safely packaged in heavy-duty cardboard for protection. The joiner then has to assemble the units before fitting them on site.

Kitchen units

Most builders today provide a fitted kitchen as part of the dwelling. Manufacturer's details can be obtained for specific examples and quality of finish, but units fall within the following basic categories:

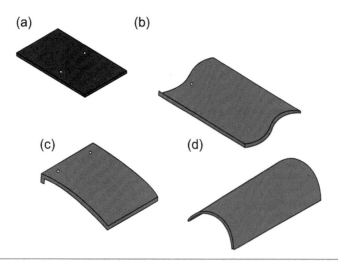

FIGURE 5.28
Roof coverings: (a) roof slate; (b) pantile; (c) standard double-lap plain tile; (d) half-round ridge tile

- base units – including provision for sinks, hobs and tables

- wall units – including upper units, cooker hoods and spot lighting

- tall units – including food cupboards, broom storage, provision for building in cookers, fridges, etc.

All are available in a variety of lengths based on a 100 mm module, in multiples of 300, 400, 500 and 600 mm. A typical assembly drawing is shown in Figure 5.29.

Safe work practices

The Health and Safety Commission has laid down certain regulations regarding the handling of materials to prevent people from being injured.

Summary of regulations

- No one must be employed to lift, carry or move any load as heavy as to be likely to cause injury.

- All employers must provide safe systems of work.

- Employers must ensure safety and absence of risks in the handling,

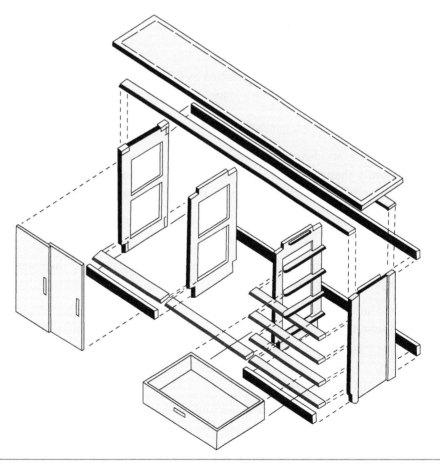

FIGURE 5.29
Flat packs

storage and transportation of all types of materials.

- Employers must provide instruction and training, to ensure the health and safety of employers.
- Employees must take reasonable care for their own safety and the safety of others.

In other words, it is the duty of the employer to ensure that you as an employee have the right training as well as the right equipment, but when actually carrying out the task of lifting or handling it is you, the employee, who must take care, not only for yourself but for others.

Common injuries

To the back/muscles

Strains and sprains to muscles, torn ligaments, back trouble and even hernias are all symptoms of the bad handling of materials.

They can be caused by sudden and awkward movement while handling heavy objects or by trying to lift an object that is far too heavy.

To the skin

Cuts and abrasions from rough surfaces, sharp or jagged edges, splinters and even projections, should be avoided.

To limbs

Limbs (arms and legs) may be crushed by falling loads, or fingers, hands or feet may become trapped by heavy loads.

Strains and twists

Most of the accidents on site are a result of employees handling heavy or awkwardly shaped objects such as cement bags, concrete and steelwork. The injury is often the cumulative result of strain that has occurred over a period.

Strains are also incurred fitting wire ropes to excavators, bar bending, pick and shovel work.

Employees also injure themselves by overreaching, twisting themselves and attempting to stop loads swinging.

Kinetic handling must be taught by skilled trainers. It cannot be 'picked up'. It is important that postural errors are corrected during training before bad habits are established.

Handling materials safely

The trainee should be able to select and use appropriate safety equipment and protective clothing when handling different materials.

Appropriate equipment and aids should be selected and used to carry materials.

The trainee should also be able to demonstrate safe manual handling techniques.

> **Note**
>
> Always use lifting gear if it is available.

Lifting gear

Numerous items of small lifting equipment are available to assist with handling materials on site and in the workshop. Only use these if you are qualified to do so.

These range from the small brick lifts, slings, barrows and dumpers to mechanical fork-lift trucks. A selection is shown in Figure 5.30.

- Barrows are the most common form of equipment for moving materials on site.
- A sack truck can be used for moving bagged materials and paving slabs.

(a)   (b)

(c)   (d)

(e)

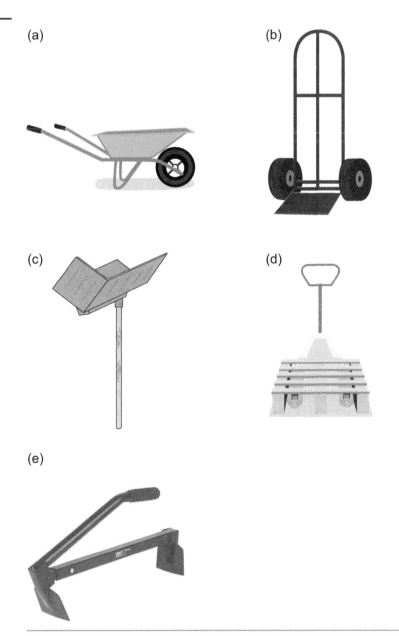

FIGURE 5.30
Various types of moving equipment: (a) barrow; (b) sack truck; (c) hod; (d) pallet truck; (e) brick tongs

- A hod can be used for moving bricks on to higher levels such as scaffolds.

- A pallet truck can be used on hard areas for moving heavy loads.

- Brick tongs are available for moving bricks and blocks around the construction site.

Many materials are delivered to the site on lorries equipped with mechanical off-loaders. Once the material has been off-loaded it is the builder's responsibility to move the materials to a secure place until required for use.

Mechanical lifting and moving machinery

Mechanical equipment

On larger sites more use will be made of mechanical equipment.

If materials are stored in a compound they will be required to be transported around the site as required.

Scaffold crane – these are useful for small jobs and are attached to the scaffold and can lift small amounts of materials to higher levels.

Dumper – there are many sizes and designs of dumper. These are only available for movement at ground level.

Forklift truck – these again are available in numerous designs and sizes and are excellent for moving materials both on the ground and at small heights.

Cranes – these are available in all makes and sizes according to the job in hand. Cranage can be either mobile or static. Cranes that span the whole construction site are excellent for the placing of all materials wet or dry.

A selection of mechanical moving equipment is shown in Figure 5.31.

Personnel protective equipment

Depending on the type of workshop or site situation, the wearing of correct safety clothing and safe working practices are the best methods of avoiding accidents or injury. On some sites certain PPE is compulsory.

All construction operatives have a responsibility to safeguard themselves and others. Making provision to protect oneself often means wearing the correct protective clothing and safety equipment.

Your employer is obliged by law to provide the following:

- suitable protective clothing for working in the rain, snow, sleet, etc.

- eye protection or eye shields for dust, sparks or chipping

- respirators to avoid breathing dangerous dust and fumes

- shelter accommodation for use when sheltering from bad weather

(a)

(b)

(c)

(d)

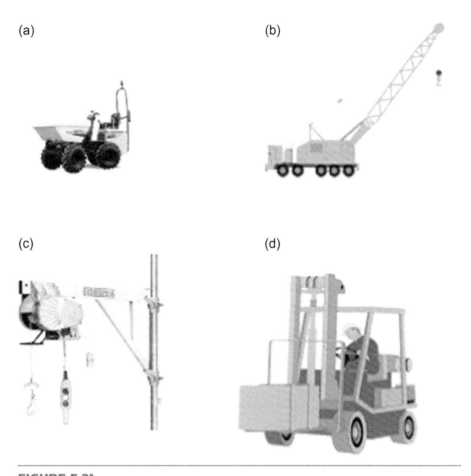

FIGURE 5.31
Types of mechanical lifting equipment

- storage accommodation for protective clothing and equipment when not in use

- ear protectors where noise levels cannot be reduced below 90 dB(A) for 8 hours.

- adequate protective clothing when exposed to high levels of lead, lead dust fumes or paint

- safety helmets for protection against falls of materials or protruding objects

- industrial gloves for handling rough abrasives, sharp and coarse materials.

Although our skin is not proof against knocks, bumps, cuts, acid, alkalis or boiling liquids, it is waterproof. Even so, we do have to cover up at times to protect ourselves.

Workers in the construction industry are liable to injury or even death if they are not protected. Because of this, protective clothing has been developed to help prevent injury.

Various items of PPE are shown in Figure 5.32.

FIGURE 5.32
Selection of PPE

Storing, stacking and protecting materials and components

Objectives of storage

The objectives of storage are to enable the quick and easy location of materials, plant and equipment.

Purpose of storage

Secure accommodation together with an effective storage system is essential on the construction site to prevent unnecessary waste and loss of materials due to vandalism and theft, and damage caused by constant handling.

Storage control

Irrespective of the amount of materials being stored, the following factors need to be considered.

Deliveries

Materials received need to be checked against suppliers' delivery advice notes for correct quantities and damage.

Stock rotation

Stocks of some materials (plaster, cement, paint, etc.) need to be used in the order of delivery as they only have a limited shelf-life.

Withdrawals

Materials drawn from stock by requisition need to be replaced.

Security

Storage facilities must be capable of being made secure against vandalism and theft.

Cleanliness

The store and unloading areas must be kept clean and free of debris at all times.

Safety

Materials must be handled and stacked in such a manner that injuries and damage are avoided. Hazardous materials need to be stored in separate accommodation constructed of non-flammable materials.

Site storage

Provisions for site storage will include the following.

Site compound

A fenced area is required for storing materials and plant, providing a means of control and protection against vandalism and theft. A typical site compound is shown in Figure 5.33.

Site stores

Site stores are needed for storing materials that must be protected against the weather and which are easily lost or stolen. Two examples are shown in Figure 5.34.

Storage areas

Predetermined areas are required on the site where materials are to be placed before use.

Materials such as bricks and blocks could be stored in the compound or placed close to where they will be used on the site, as shown in Figure 5.33.

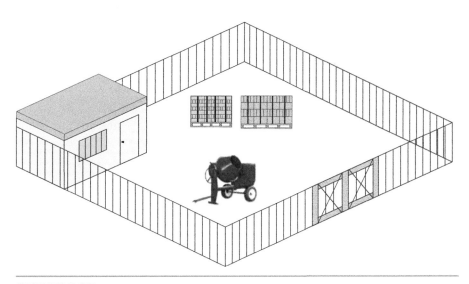

FIGURE 5.33
Typical site compound

(a)   (b)

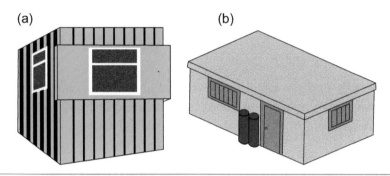

FIGURE 5.34
Types of material store: (a) metal; (b) wooden

Workshop/yard storage

Convenient workshop storage should be provided for tools and equipment.

Storage under cover should be provided for timber, sheet and insulation materials.

Separate workshop accommodation may be required for the storage of paint and inflammable fuel.

A maintenance area should also be provided so that plant and equipment can be maintained and cleaned before being stored or sent out to other sites.

Material storage

The amount of storage on site depends on the size of the contract and the amount of space available for storage.

On a large site storage facilities can be very comprehensive, but on small sites the materials are often delivered as required, materials being retained at the builder's merchant until called for.

Some firms will store small items at their main depot and the site supervisor will call them forward with requisition sheets.

Even on large sites some materials are delivered as required, such as ready-mixed concrete, mortar and hardcore.

Figure 5.35 shows a typical site for a factory unit. Areas have been set out for material storage and security. The whole site has a fence around it and an internal compound.

Aggregates

Aggregates should always be kept in a clean condition and should not become contaminated with undesirable substances.

Not only can soil, clay, organic or deleterious materials impair the strength of the concrete, but they can also stain the mix, resulting in patchiness or an unsightly finish.

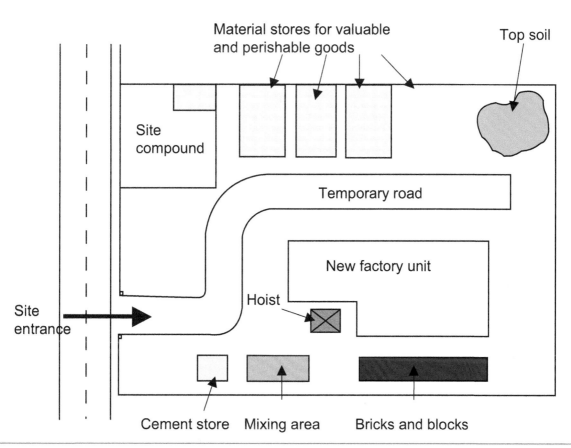

FIGURE 5.35
Typical site layout for materials

Aggregates stored in builder's yards are usually stored in specially prepared areas.

Figure 5.36 shows a typical area for storing aggregates on site. These areas are usually built with a concrete slab, sloping to the outside, and brick walls separating the various types of aggregates and sands.

On the site aggregates are very often stored directly on the ground. In this case the top soil should be removed and the pile should be as large as possible, with the bottom 500 mm never being used.

Aggregates and sands that have become contaminated should be removed from site.

In severe weather the stockpiles should be protected by sheeting materials.

Both very hot and cold materials will reduce the quality of the finished mix.

Bagged materials

These should be stored in a weatherproof shed, preferably without windows, and raised off the ground. An example is shown in Figure 5.37.

Cements, plasters and lime are the main bagged materials, all of which will deteriorate rapidly when damp.

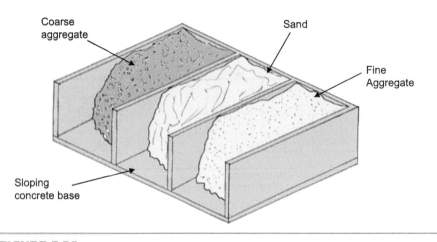

FIGURE 5.36
Site storage for aggregates

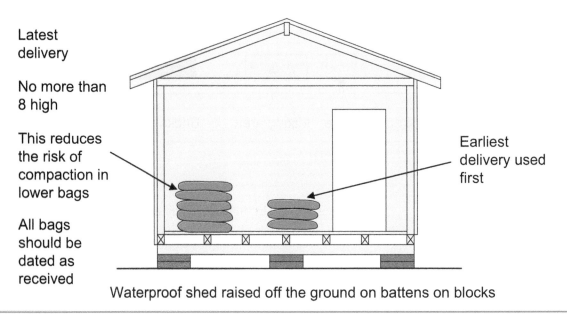

FIGURE 5.37
Cement storage

The materials should be stored separately and clearly dated.

They must be stored in such a way that they can be used on a first-in, first-out rotation. This is to minimize the storage time and prevent the bag contents becoming stale or 'air setting'.

Care should be taken when using bagged materials as the paper can cause a problem on the site.

Bricks, blocks and stone

Bricks, blocks and stone should be stacked clear of the ground and covered completely with tarpaulins (Figure 5.38).

Tarpaulins used to cover materials

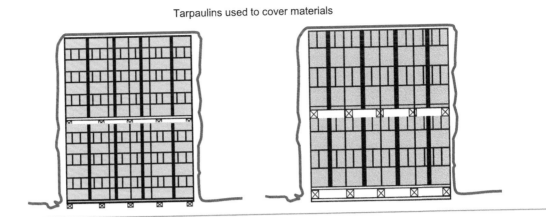

FIGURE 5.38
Protection of materials

Some manufacturers deliver their materials already shrink-wrapped. The wrapping should be kept on until the materials are ready for use.

Drainage materials

Drain pipes should be stored in the open where they will not be accidentally damaged, on firm, level ground that has been covered with hardcore (Figure 5.39).

Because of their shape they must be restrained from rolling about and occupying a large ground area.

If the drain pipes have socketed ends they should be stacked with sockets at alternative ends for protection.

Concrete goods

Most concrete goods can be stored outside on a level, well-drained site. Examples are shown in Figure 5.40.

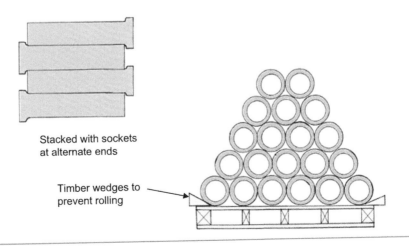

Stacked with sockets
at alternate ends

Timber wedges to
prevent rolling

FIGURE 5.39
Stacking of drain pipes

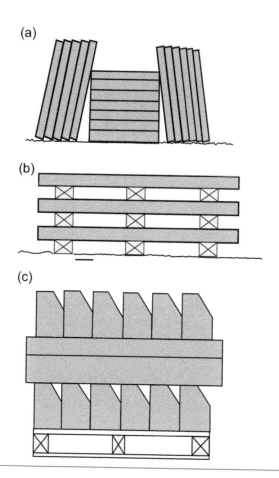

FIGURE 5.40
Storage of concrete goods: (a) paving slabs; (b) lintels; (c) kerbs

Paving slabs should be stored on their edge, standing on timber bearers. Different sizes, shapes and colours should be stored separately.

Concrete lintels should be stored on timber bearers, but it is essential to ensure that they are stored in the way they are to be fixed, i.e. top side uppermost.

Kerbs should be stored on edge on timber bearers. They may be stored in bonded layers, but never more than three high.

Roofing tiles

Roofing tiles and slates should be stored on level, well-drained ground, possibly stoned to keep the area clean. See the examples in Figure 5.41.

Care should be taken when storing roofing tiles and slates to avoid damage.

Roofing tiles are impervious to moisture and can be stored outside. However, on a building site they should be wrapped around with a tarpaulin to protect them from splashing.

Plain tiles are stacked on end, end tiles flat and taper stacked. The tiles are placed alternately to prevent damage to the nibs. Ridge tiles are stacked as shown – never more than 10 high.

(a)   (b)

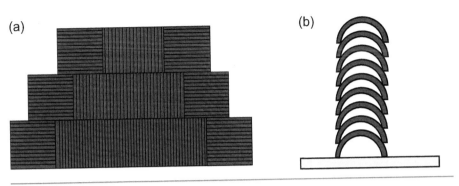

FIGURE 5.41
Stacking of roof coverings: (a) plain tiles; (b) ridge tiles

Ceramic tiles

These are delivered to the site in boxes and after checking for damage should be stored in their boxes (Figure 5.42). They should be stored inside a store or the building where they are required.

Sheet materials

Plywood

Small quantities of plywood are best stored in racks, while large stocks can be stacked on bearers spaced at 400 mm centres to prevent distortion and covered with a tarpaulin sheet (Figure 5.43).

When handling care should be taken to avoid damage to corners and edges.

Plasterboard

Plasterboard should be stored in a similar way to other sheeting materials.

FIGURE 5.42
Storage of ceramics

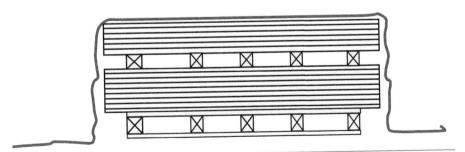

FIGURE 5.43
Storage and protection of sheet materials

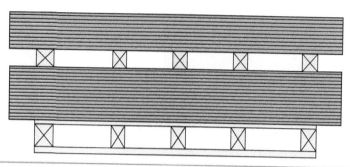

FIGURE 5.44
Storage of plasterboards

Plasterboard can be unloaded directly and stored in the building where it is required, laid flat on timber bearers (Figure 5.44).

Timber lengths

Poor storage of timber, wood products and units is a major cause of material wastage on building sites. If carcassing timber has to be stored outside then it should be stored off the ground on timber battens and protected against the weather with a tarpaulin cover (Figure 5.45).

It is extremely important to ensure that the moisture content of the timber is kept at the right level for its use.

Timber components are usually delivered to site when required and should be stored inside a large, dry store or using part of the completed building as a store.

Always lay them flat and never lean them against a wall, etc., or they will bow, making them unusable (Figure 5.46).

Paints and flammable liquids

Paints should be stored inside a special store as shown in Figure 5.47.

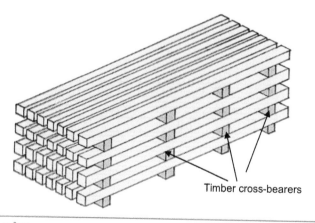

Timber cross-bearers

FIGURE 5.45
Storage of timber lengths

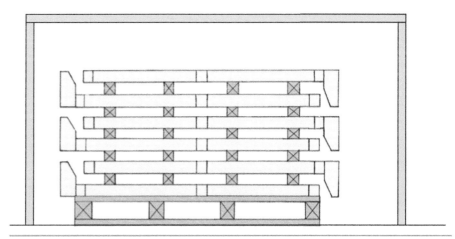

FIGURE 5.46
Storage of timber components laid flat under cover

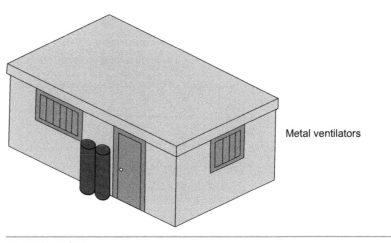

Metal ventilators

FIGURE 5.47
Storage of flammable liquids

Flammable liquids should be kept in securely capped cans or steel drums on which the contents are clearly marked.

Petrol, acetone, methylated spirits and other volatile liquids with flashpoints below 32°C may be kept in robust metal lockers or well-ventilated non-combustible cabins, or in the open.

Storage should be securely locked or fenced, and be situated at least 6 m away from any other huts, stores, site boundaries or buildings.

Smoking should be prohibited within 6 m of the store.

No other materials should be stored with flammable liquids.

Containerized items

Any items delivered to the construction site in cardboard boxes should be ordered when required and stored in waterproof, dry conditions (Figure 5.48).

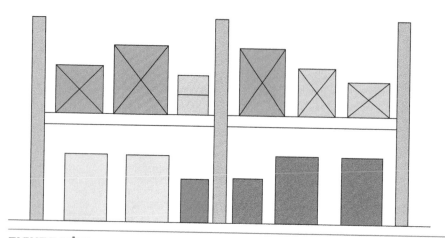

FIGURE 5.48
Storage of containerized materials

A secure store should be provided unless there are sections of the building that are complete and lockable which could be used.

Paint is delivered in tins, which also needs special arrangements for storage. Paint has a limited shelf-life so should only be ordered as and when required.

Shelving should be provided in the store so that items can be easily stored and retrieved.

Rolled materials

All rolled materials should be stored in an appropriate store, on end to retain their shape (Figure 5.49). They should be stored in their appropriate sizes for ease of identification.

They should be stored away from any source of heat as they could melt and become useless.

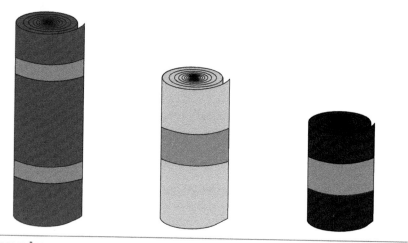

FIGURE 5.49
Storage of rolled materials

Minimizing damage and waste

NEVER stack DPCs flat as this causes the rolls to flatten and crack.

Roofing felt should be stacked in a similar fashion, on end and in special types for ease of identification.

Rolls of lead should also be stored on end in their separate codes. It is important to protect lead from damage, as damaged lead is useless. Lead should kept be in a lockable shed or building.

With the present trend of material costs rising at a faster rate than labour costs, the control of waste is very important both on and off site.

The design of the building can add to the waste problem by involving cutting of materials.

Waste can be divided into:

- waste from mixing and cutting
- waste from poorly stored and protected materials
- waste from overordering material supplies
- waste from theft
- waste from the nature of the application of the material, e.g. mortar dropped on the floor by the bricklayer.

This chapter will only deal with the management of the resources on site.

As described previously, materials can be delivered to the site in various ways and the site manager is responsible for the best handling and storage to protect the materials from damage.

Materials should be ordered in exact quantities. In some cases this may not be economical owing to transport costs imposed by the supplier.

Loose materials such as building sand and ballast are ordered by the load but invoiced by the tonne.

Only order the required amount, allowing only 5 per cent for wastage, as the remainder is always left behind. A central mixing area, or using ready-mixed materials, can help to minimize wastage in this area.

Bagged materials are also a source of wastage, especially if part-open bags are left out at night and get damp.

The storage of materials on site is the direct responsibility of the site management and has a very strong influence in the control of waste. A careless regard for the value and utility of the materials on the part of management could lead to a progressive deterioration in the operatives' regard for the material.

Areas allocated on site for the storage of materials should be determined after considering the following questions:

- Will construction take place in that area? If so, when?
- Is the storage for a long or a short time?
- Can delivery vehicles safely and easily reach the area?
- Can on-site movement from storage areas to point of use be safely and economically carried out?
- Are the materials as near to their point of use as practically possible?
- If the materials have considerable value are they in the secure area?

- Will the storage area create problems in routing site transport and personnel?

The answers to the above questions will, of course, vary from site to site.

Please refer back to Figure 5.35, which shows a typical site layout with areas set out for material storage.

Disposal of materials from heights

Careful disposal of materials from heights is essential.

They should always be lowered safely and never thrown or dropped from scaffolds and window openings, etc.

Chutes can be employed as shown in Figure 5.50. These can be connected together to form any reasonable length. They should terminate in a tarpaulin-covered skip.

Waste can be kept in a skip until collected and taken to a suitable tipping area.

Hazardous waste, non-hazardous waste and debris

Waste resulting from work materials can include sand, gravel, cement, concrete, bituminous waste, sealants, bricks, stone, steel mesh, nails, timbers, fuels, oils, etc.

Waste resulting from work activities can include soils, vegetation, cuttings, inorganic materials, etc.

Depending on the sector in which the operative works, the materials will differ. Some operatives will work in a specific trade occupation, relating to bricks or

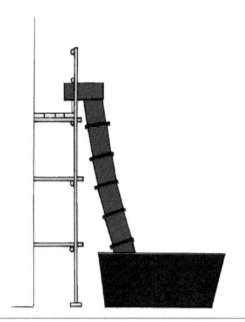

FIGURE 5.50
Removing waste from a building

timber, while others will be involved in general construction activities, relating to excavation materials.

Most substances and materials are safe provided they are handled or worked on sensibly and with proper precautions. Almost anything can be dangerous if handled or used irresponsibly. Some materials require extra care; a few need extreme caution.

It is important to know what the hazards are, when they occur and how they can be prevented. Usually prevention will involve the use of protective clothing and equipment.

Removal of waste

Since 1 April 1992, the law on waste has included a duty of care that means you must take all reasonable steps to look after any waste you have and prevent its illegal disposal by others.

If you break the law, you could be fined an unlimited amount.

'Controlled waste' is any household, commercial or industrial waste, such as waste from a house, shop, office, factory, building site or any other business premises.

How do I know what to do?

The law requires you to complete certain paperwork and to take all reasonable steps to meet the duty of care. What is reasonable depends on who you are and on the circumstances. *The Duty of Care, A Code of Practice*, is published by HMSO (ISBN 011 75 25 57 X).

Steps to be taken if the duty of care affects you

- When you have waste you have a duty to stop it escaping. Store it safely and securely.
- If you hand the waste to someone else:
    1. Secure it.
        - Most waste should be in suitable containers.
        - Loose material loaded into a vehicle or skip should be covered.

    2. Check that the person taking your waste away is legally authorized to do so.

Multiple-choice questions

Self-assessment

This section of the book is designed to allow you to check your level of knowledge. The section consists of revision questions for this chapter. The questions are all multiple choice and have four possible answers. The answers are to be found at the end of the book.

The main type of multiple-choice question will be the four-option multiple-choice question. This will consist of a question or statement, known as the stem, followed by a choice of four different answers, called the responses. Only one of these responses is the correct answer; the others are incorrect and are known as distracters.

You should attempt to answer the questions by choosing either (a), (b), (c) or (d).

Example

The person employed by the local authority to ensure that the Building Regulations are observed is called the:

    (a) clerk of works

    (b) building control officer

    (c) council inspector

    (d) safety officer

The correct answer is the building control officer, and therefore (b) would be the correct response.

Moving and handling resources

Question 1 Which of the following would be classed as bulk materials?

    (a) window frames

    (b) aggregates

    (c) plasterboards

    (d) ironmongery

Question 2 Lengths of timber should be stored:

    (a) in the open

    (b) leaning against a wall

    (c) on timber bearers and covered with a tarpaulin

    (d) on the ground and covered with a tarpaulin

Question 3 Bagged material should not be stored more than:

    (a) 2 bags high

    (b) 4 bags high

    (c) 6 bags high

    (d) 8 bags high

Question 4 Which of the following is a natural material?

    (a) timber

    (b) brick

    (c) block

    (d) steel

Question 5 Which of the following is best suited for carrying bricks up a scaffold?

    (a) hod

    (b) barrow

    (c) pallet truck

    (d) sack truck

Question 6 Which of the following is classed as perishable goods?

    (a) bricks

    (b) building sand

    (c) plasterboards

    (d) blocks

Question 7 Which of the following is the best method for storing roofing felt?

    (a) laid flat in a store

    (b) laid flat on ground

    (c) on end on the ground

    (d) on end in a store

Question 8 Identify the following type of brick:

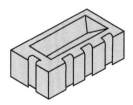

    (a) solid wire-cut

    (b) pressed or hand-moulded

    (c) cellular

    (d) keyed

CHAPTER 6

Solid Walls and Piers

This chapter will cover the following City and Guilds units:

- Apprenticeship   Module 5   S1.1, S1.2, S1.3, S3.1, S4.2, S4.1, S4.2, S5.1, S5.4, S5.1
                                   K1.1, K1.2, K1.3, K3.1, K3.2, K3.3, K3.4, K4.1, K5.1
                   Module 12   S3.1, S3.2, S4.1, S5.1, S5.2, S6.1
- Certificate   Unit 203   1.1, 1.2, 1.3, 2.1, 2.2, 2.3, 2.4, 3.1, 3.2, 4.1, 4.2
- Diploma   Unit 201   1.4, 1.5, 1.6
                                   2.4, 2.5, 2.6, 2.7, 2.9, 2.10, 2.11
                                   3.1, 3.2, 3.3, 3.4, 3.5, 3.6, 3.7, 3.8, 3.9, 3.10, 3.11, 3.12, 3.13
                                   4.1, 4.2, 4.3, 4.4, 4.5, 4.6, 4.7, 4.8, 4.9, 4.10, 4.11, 4.12, 4.13, 4.14
                                   5.1, 5.2, 5.3, 5.4, 5.5, 5.6, 5.7, 5.8, 5.9, 5.10, 5.11, 5.12, 5.14
                                   6.1, 6.2, 6.3, 6.4, 6.5, 6.6, 6.7, 6.8, 6.9, 6.10, 6.11, 6.12, 6.13, 6.14

This chapter is about:
- Interpreting building information
- Adopting safe and healthy working practices
- Selecting materials and components and equipment
- Preparing and erecting solid walling

The following Skills and Knowledge requirements will be covered:
- Understand and build solid walls and isolated and attached piers
- Apply finishes to top of walls and piers

The following Certificate learning outcomes will be covered:
- Plan, prepare and build solid walls and piers
- Apply finishes to tops of walls and piers

The following Diploma learning outcomes will be covered:
- Know how to plan and erect solid walls and piers
- Know how to finish tops of walls and piers

Interpretation of information

Sources of information

There are many items of information available to the trainee bricklayer with regard to solid walls and their application.

Many builders have numerous sources of information such as:

- manufacturers' literature
- specifications
- Building Regulations
- British Standards Institution
- British Standard specifications
- codes of practice
- Building Research Establishment (BRE)
- BRE digests, information papers and current papers.

The trainee can also visit libraries to find information from:

- books and technical articles.

Many libraries also take trade periodicals which operate a reader's enquiry service whereby technical brochures and information can be obtained from various manufacturers and suppliers.

Trade periodicals:

- *Cement and Concrete Monthly*
- *The Architects' Journal*
- *Building*
- *The Brick Bulletin*

and many more.

When information has been received it is important that the trainee can interpret it and be able to use it correctly.

Other information can be received from supervisors and workmates by oral or written methods.

Remember there is no real proof that the person has received the correct information, but a quick response can be achieved with oral instructions.

This chapter is also concerned with the selection, maintenance and safe use of power tools and equipment.

Whether dealing with powered static equipment or hand-held powered tools, or carrying out predictable and routine operations with powered tools and equipment and designated small mobile plant, identifies good practice procedures that need to be habitual to the work operator.

Before selecting the power tools for use on a contract it is essential to know all about the various tools and equipment, especially the powered ones.

Technical information can be obtained in a variety of ways and from a variety of sources. Most tools and equipment come with manufacturer's safety instructions and technical information.

Technical information can be produced in several formats:

- operating instructions – how to use the item

- safety guidelines – power supply, personal protective equipment (PPE) to be worn and recommended checks

- technical information – mechanical details and possible outputs.

Information

Normal hand tools are not usually provided with manufacturer's instructions.

Most textbooks explain hand tools in detail and the safety procedures to follow while using them.

Information on hand-held power tools may be easy to find and is usually received along with the purchase of the particular item of tool or equipment.

This may cause a problem if the item is being hired for a period, but the hirer should be able to supply technical, safety or operating instructions.

Safe work practices

Abrasive wheels/stone saws

The most common type of power tool in use by the bricklayer when building solid walls is the angle grinder or disc cutter.

These tools are used for cutting various types of materials from stone and brick to concrete and tarmac.

They are available in all types and sizes, ranging from small hand-held skill saws to bench-operated disc cutters (Figure 6.1).

Cutting with power tools

When components are cut by machine it is essential that the equipment is set up correctly and safely and the operator has had sufficient training and has acquired the required certificate.

As for hand cutting, the appropriate PPE must be worn.

The main type of hand-held equipment for cutting building components is the angle grinder or disc cutter. A table version of a disc cutter is used in workshops and building sites when a great deal of cutting has to be done.

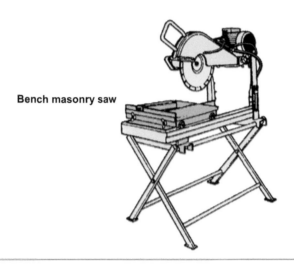

Bench masonry saw

FIGURE 6.1
Cutting equipment

Portable power tools

Numerous portable power tools are now available for all trades and it is essential that the manufacturer's instructions are read and understood before setting up and using the equipment.

Whenever power tools are being used on the construction site it is essential that the correct safety procedures are adhered to at all times.

Please refer back to Chapter 2, Health and Safety in the Construction Industry, as required.

Electricity

You will need a basic understanding of:

- the dangers:
  - what makes electricity so dangerous
  - what electric shock can do to the body

DO NOT CHANGE
GRINDING WHEEL
UNLESS AUTHORIZED
TO DO SO

FIGURE 6.2
Abrasive wheel warning notice

- what type of work might bring you near to live electrical cables.

- avoiding danger:

  - how to plan your work so that you are not put at risk

  - what precautions to take so that you and others are not put at risk of electrocution

  - how to avoid injury from overhead cables

  - how to avoid injury from buried cables

  - how to recognize danger areas into which you should not go

  - the maximum safe voltage for use on site

  - how to reduce the voltage to a safe level.

Electricity supply

The supply of electricity will normally be provided by one or both of the following:

- public supply from the local electricity board

- a site generator when public supply is not practicable or uneconomic.

Site distribution

The electricity supply for operatives to use is distributed around the site using a number of different units.

All supply, distribution and transformer units should be marked with the warning sign shown in Figure 6.3. An additional sign should be placed under the warning sign and should state the highest voltage likely to be present.

Plugs, sockets and outlets and couplers

To avoid plugs designed for one voltage being connected to sockets of another voltage, there are different positions for the keyway in plugs and sockets. The examples shown in Figure 6.4 conform to British Standard BS 4343.

DANGER 415 V

FIGURE 6.3
Electric safety sign

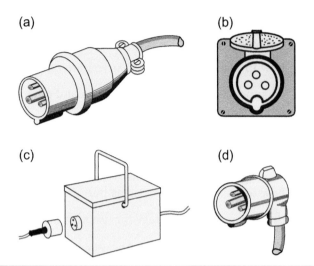

FIGURE 6.4
110 V plugs and connectors

Cables and wires

In all site offices, worksites and workshops and similar premises wiring should comply with existing Wiring Regulations.

Since 2004 all low-voltage cable colours have had to comply with the cable colours of the European Union (Figure 6.5).

These colours are:

- neutral – blue (previously black)
- earth – green/yellow (same as previously)
- phase 1 – brown (previously red)
- phase 2 – black (previously yellow)
- phase 3 – grey (previously blue).

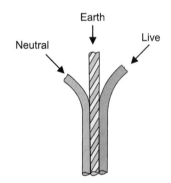

FIGURE 6.5
Wiring colours

Wiring for plugs

A wiring diagram for an electrical three-pin plug in everyday use is shown in Figure 6.6.

Electric circuit breakers

The residual current device (RCD) (Figure 6.7) is also known as an earth leakage circuit breaker (ELCB) or residual current circuit breaker (RCCB).

This indispensable device cuts off all current in a fraction of a second if it defects to a sudden leakage to earth due to a fault.

Always use one when working with electric power tools.

Warning

NEVER overload sockets with adapters.

NEVER use a cable that has not been fully unwound.

Selection of resources

When building solid one-brick walls the resources will simply be tools, bricks and mortar.

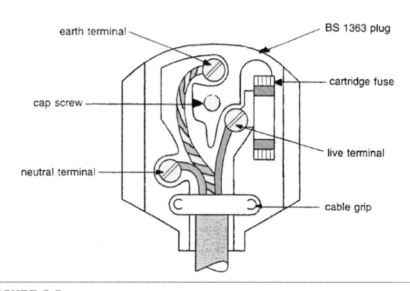

FIGURE 6.6
Wiring diagram for an electrical three-pin plug

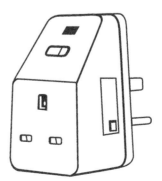

FIGURE 6.7
RCD breaker

This chapter will assume that the work has been completed below ground level, so the first procedure will be setting out the materials at ground level.

Setting out materials

Before brickwork commences the working area must be stacked out with bricks and blocks as in Figure 6.8.

Keep stacks away from edges trenches. Keep walkways and runways clear. Do not stack materials too high. Do not stack wider at the top than at the bottom.

Stack materials near the job to avoid double handling. Keep materials at working height.

Protect materials from breakages, dampness and frost. Cover bricks with tarpaulin, hessian or polythene.

Take care when opening packs of bricks. The polythene wrapping or banding should be left in place until the bricks or blocks are required. The correct way to open packs is by cutting the straps with snips. Do not chop the straps with a brick hammer or trowel or any tool that will damage the bricks. For safety reasons, take the banding to the skip immediately, or make into safe bundles and remove from the work area later.

Polythene wrapping, if removed carefully and placed to one side, could be used elsewhere for protection. If it is not required remove to the skip as soon as possible.

When loading out for the bricklayer place the bricks with the frog or perforations uppermost, for easy handling by the bricklayer.

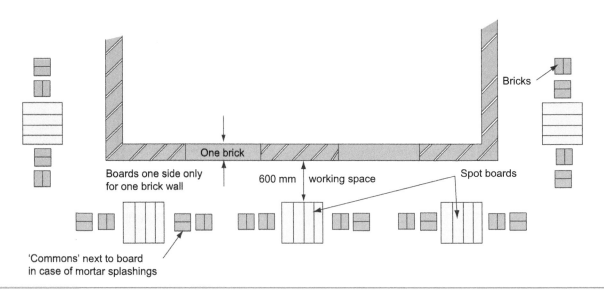

FIGURE 6.8
Setting out materials before building

Source the bricks from as many packs as possible, but at least three. This will prevent banding along the wall of different coloured or textured bricks. Banding of bricks can occur in the face of walls if bricks are taken from only one pack at a time (Figure 6.9).

Minimizing risks

It is important that the trainee is aware of the dangers on site and takes the necessary precautions such as wearing the appropriate PPE.

One of the most dangerous aspects of solid brick walling, power tools, has already been covered. Another important task to watch is the movement of materials around the site.

Transporting bricks

Bricks can be carried under the arm from the stack to where they are required, but this is laborious and slow.

The use of brick clamps which can pick up six bricks on edge saves the hands and is quicker.

Barrows are much quicker, although barrow runs will have to be laid over rough ground.

Brick trolleys which carry 48 bricks held by a clamp are best used on the scaffold or on smooth ground. Bricks must be stacked on edge ready for the brick trolley.

The most common method used for moving bricks on larger sites is by fork-lift truck (Figure 6.10).

Protection of the surrounding area is important so as to minimize risks to the public.

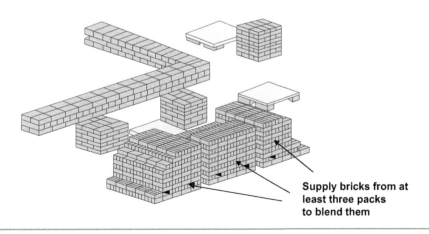

Supply bricks from at least three packs to blend them

FIGURE 6.9
Selecting bricks when setting out

FIGURE 6.10
Moving bricks with the fork-lift truck

The building site should be kept clean and free from as much dust and noise as possible. All waste materials should be collected in a skip, which can be moved off site when full.

The new walling should be protected with sheeting as the work proceeds, until fully set.

Meeting the contract specification

Bonding solid walls – recap on basic rules

Bonding of brickwork

The bonding of brickwork is required for its appearance as well as its strength.

- a bond used for strength – English bond
- a bond used for appearance – Flemish bond.

Bonds can also be used for their economical use of facings, such as the garden wall bonds.

Table 6.1 shows the number of facing bricks required according to the type of bond.

Principles of bonding

To maintain strength, bricks must be lapped one over the other in successive courses along the wall and in its thickness.

There are two practical methods, using either a half-brick lap or a quarter-brick lap, called half-bond and quarter-bond (Figure 6.11).

If the lap is greater or smaller than these, then both appearance and strength are affected. If bricks are so placed that no lap occurs, then the cross-joints or perpends are directly over each other (Figure 6.12), and this is termed a straight joint, being either external for those appearing on the face of the wall or internal for those occurring inside the wall, and they should be avoided whenever possible.

Table 6.1 Number of facing bricks required according to bond (per 0.84 m2)			
Type of bond	Thickness of the brick		
	50 mm	65 mm	70 mm
English bond	90	72	66
Flemish bond	80	64	59
Dutch bond	90	72	66
English garden wall bond	75	60	55
Flemish garden wall bond	69	56	51
Header bond	120	96	88
Stretcher bond	60	48	44

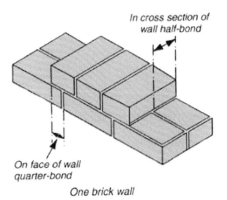

In cross section of wall half-bond

On face of wall quarter-bond

One brick wall

FIGURE 6.11
No straight joints

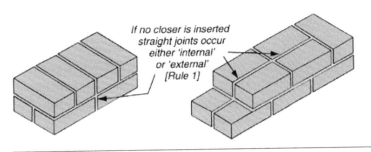

If no closer is inserted straight joints occur either 'internal' or 'external' [Rule 1]

FIGURE 6.12
Straight joints

The apprentice should note that internal straight joints will occur in some bonding problems. On the one hand, excessive cutting may solve a particular problem, but this wastes labour and materials and tends to weaken the wall. On the other hand, by introducing one or two straight internal joints, whole bricks can be used. This is a case where practice and theory must compromise.

The pattern in a brick wall is purposely arranged, has its particular use, and is called a bond.

To summarize, the two main principles of the bonding of brickwork are:

- to maintain half- or quarter-bond, avoiding external straight joints and internal straight joints wherever possible
- to show the maximum amount of specified face bond pattern possible.

To assist in maintaining these principles, rules should be remembered and applied (Table 6.2).

The apprentice should never try to remember all the problems shown as examples. Problems must be solved as they occur by the logical application of the rules. Eventually, the bonding of brickwork becomes automatic to the bricklayer.

Several bonds are in general use, but for the purpose of beginning the apprentice's bonding education, stretcher, English, and Flemish bonds will be explained. Problems in other bonds can be solved by the application of the same rules.

Dimensions

The bricklayer should remember the various dimensions to which bonds work without having to resort to broken or reverse bond.

Table 6.3 shows multiples of brick sizes, using normal size bricks of 215 mm, not including a 10 mm joint.

Setting out facework in a wall without openings

Before setting out the bond on the face side of any wall, it is wise to make a gauge rod (Figure 6.13). This should be made of straight, smooth timber 50 × 50 mm in cross-section, and approximately 3 m long. Along one face fine saw cuts carefully made at 225 mm intervals will allow stretcher bond to be set out consistently from end to end, leaving any broken bond near the middle of the wall.

On another surface of the rod, saw cuts at 75 mm intervals serve to check regular vertical gauge when bricklaying commences and quoins are raised. Use of this dual-purpose storey rod, for checking horizontal and vertical gauge, avoids the potential risk of error when using a measuring tape.

Setting out bonds

When the brickwork of a building approaches ground level it is normal practice to change from common bricks to facing bricks. This changeover normally takes place two courses below finished ground level to avoid commons being shown when the building is complete.

When the facings reach ground level preparation must be made in the bond arrangement for inserting door and window openings at a future height. It is at this stage that the proposed elevation of the building must be carefully studied.

This chapter will deal only with walls without openings.

> **Note**
>
> Wall thicknesses are usually stated in brick sizes, e.g. the width of a brick is known as a half-brick wall; the length of a brick as a one-brick wall; the width, plus length of a brick, as a one-and-a-half brick wall; and so on.

Table 6.2   Rules of bonding

Rule 1	The LAP of brickwork along the face of a wall shall be ¼ brick 56.25 mm	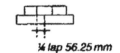
	Exception: Half-lap of 112.5 mm in stretcher bond	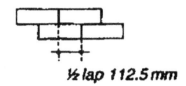
Rule 2	The LAP of brickwork along the face of a wall shall be 112.5 mm	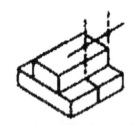
Rule 3	Brickwork should be set out on the face side and from each end of a wall or pier, so that any BROKEN BOND is centrally located	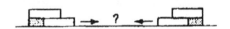
	Exception: Reverse bond, where end bricks do not correspond	
Rule 4	In English and Flemish bonds a queen closer must be placed next to the QUOIN HEADER, in order to establish ¼ lap	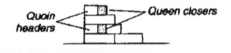
	Exception: It is permissible to use a ¼ brick instead in Flemish bond, if this will avoid broken bond at centre, or is otherwise preferred	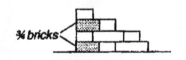
Rule 5	Cross-joints in consecutive courses of bricks must not coincide one above the other to create STRAIGHT JOINTS	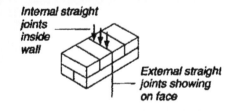

(continued on next page)

Table 6.2 (*continued*)

Rule 6	Cross-joints in alternate courses must coincide vertically one above the other on the face of the wall (PERPENDS)	

Perpends

Rule 7	The TIE-BRICK between one brick thick English bond walls is always a HEADER

Rule 8	The TIE-BRICK between one brick thick Flemish bond walls is always a STRETCHER

Rule 9	In English bond, when a wall changes direction, the face bond changes from headers to stretchers or vice versa in the same course

Exception: This rule will not apply when (a) walls of different thickness interact, (b) walling curves on plan, (c) where the change of direction is 112.5 mm or less

Exception

(*continued on next page*)

Table 6.2 (*continued*)

Rule 10	When bonding intersecting walls, run the bonding of each wall across the other in alternate courses

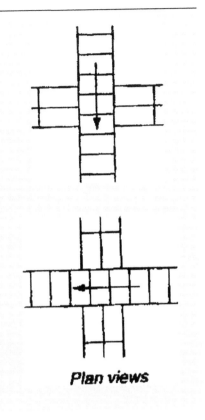

Plan views

Table 6.3	Multiples of brick sizes		
No. of bricks	Brick size	Joint size	Total length
1	215	–	215 [a]
1½	215 + 107.5	10	332.5
2	430	10	440
2½	537.5	20	557.5
3	645	20	665
3½	752.5	30	782.5
4	860	30	890
4½	967.5	40	1007.5
5	1075	40	1115
5½	1182.5	50	1232.5
6	1290	50	1340
6½	1397.5	60	1457.5
7	1505	60	1565
7½	1612.5	70	1682.5
8	1720	70	1790
8½	1827.5	80	1907.5
9	1935	80	2015
9½	2042.5	90	2132.5
10	2150	90	2240

Notes
[a] Not practicable.

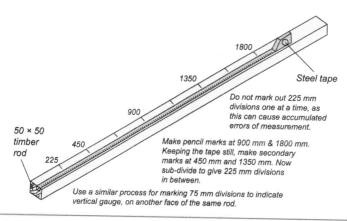

FIGURE 6.13
Marking out a gauge or storey rod

Dry bonding

The normal practice is to dry bond the bricks from each end of the wall, placing the 'broken bond' in the centre of the wall.

Setting out facework is usually the responsibility of the supervisor who will, after consulting the architect, determine the detailed bond pattern and the location, if any, of broken or reverse bonds.

Setting out the brickwork is completely different from setting out the building, which is done before excavation begins.

It is essential that no straight joints occur.

Reveal bricks provide fixed points between which the bonding is set out.

On no account should a closer be built in the middle of the wall; it should only be placed next to the quoin header.

Quoins

Quoins occur at the end of a straight wall, known as a stopped end (Figure 6.14).

Returns occur when a wall is being continued around the corner (Figure 6.15). The corner should be erected, continuing from the work completed up to ground level.

> **Remember**
>
> Change direction – change bond.

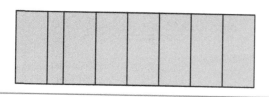

FIGURE 6.14
One brick stopped end

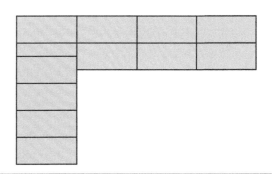

FIGURE 6.15
One brick quoin

The corners should not be erected too high as it is much more economical to run in the wall with the aid of a line than to erect large corners and rake them back to form corners.

Toothing should be avoided where possible. Some clerks of works may not allow toothing at corners as it could cause a weakness if the joints are not filled correctly. See an example of raking back in Figure 6.16. Toothing the corner produces a weakness when the wall is built into it if the joints are not con-solidated (Figure 6.17).

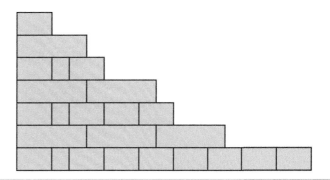

FIGURE 6.16
Raking back produces the strongest corner

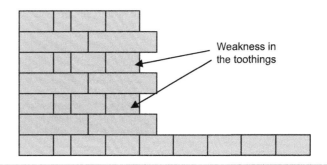

Weakness in
the toothings

FIGURE 6.17
Toothing a corner

Damp-proof course

Brickwork above ground level usually commences with the damp-proof course (DPC) (Figure 6.18).

Engineering bricks are often used for DPCs and the common type is known as a 'blue brick' DPC. This is usually two courses laid in a mortar equal to the strength of the engineering bricks, which is 1:3.

Engineering bricks are difficult to lay as they can 'swim' in the mortar owing to their high density and lack of absorption. It is therefore essential to keep the bricks as dry as possible and the mortar as stiff as possible to allow easy spreading.

Two courses of DPC bricks bedded in cement mortar provide a better resistance to overturning of walls than flexible DPCs.

This will be reinforced when boundary walls and piers are dealt with later.

Note

Do not attempt to joint the bricks until the mortar has hardened sufficiently.

Tingles

When building walls longer than 1.125 m long it is always advisable to use lines and pins.

If the wall exceeds 9 m in length then it is advisable to use a tingle plate to prevent the line from sagging (Figure 6.19). The tingle brick should be as near as possible to the middle of the wall.

Collar joint

This is the joint that runs along the centre of the stretcher course when building a wall in English bond (Figure 6.20).

Different skills are required when building one-brick walls.

Stretcher course

Care should be taken when laying the stretcher course as too much mortar between the bricks will cause the face of the wall or the back of the wall to bend.

The face of the wall should be laid first to the line provided.

When the mortar is laid for backing the wall it should be pulled back from the face brick so that when the backing brick is laid it does not push out the facing brick (Figure 6.21).

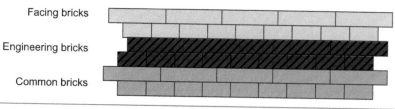

Facing bricks

Engineering bricks

Common bricks

FIGURE 6.18

Damp-proof course in engineering bricks

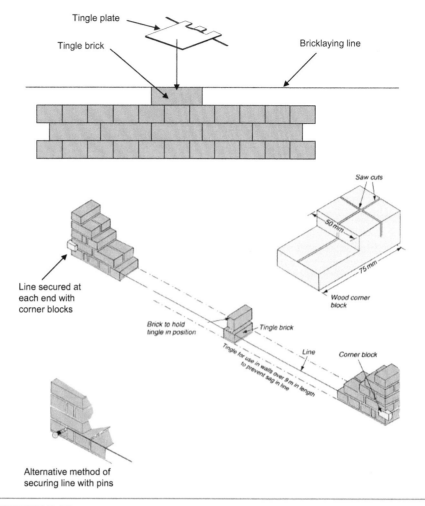

FIGURE 6.19
Use of tingle plate to prevent line sagging

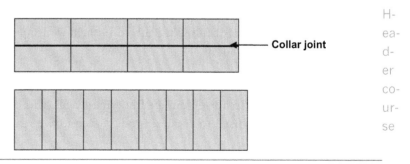

FIGURE 6.20
Alternate courses in English bond

When the header course is being laid special attention should be paid to the long cross- joint (sectional joint) along the stretcher face of the brick (Figure 6.22).

The brick will have to be turned in the hand to expose the longer face of the brick.

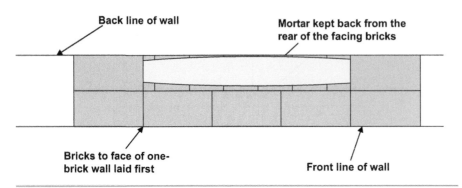

FIGURE 6.21
Laying back stretchers

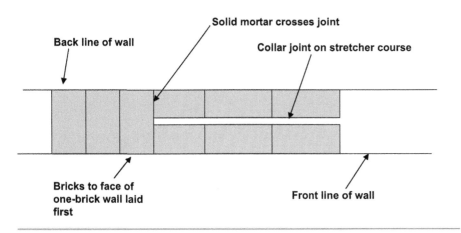

FIGURE 6.22
Laying the header course

The cross-joint should be applied in full, not top and bottom, to produce a solid joint.

The correct action is vital if full joints are to be achieved. Again, practice is necessary to achieve full joints.

Laying the back course of stretchers.

It is usual to lay the back course from the front of the wall. This is another extra skill that has to be practised to obtain a straight face to the rear of the wall.

Perpends

There are more cross-joints when building one-brick walls.

It is essential that they are controlled and that perpends are kept plumb. This can be done by checking with a spirit level or using a T square (Figure 6.23).

> **Remember**
>
> There is no line to the back of th[e] wall so it is essential to use the ey[e] to line the bricks in with the cours[e] below.

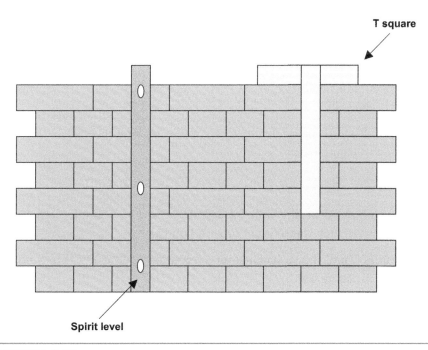

FIGURE 6.23
Plumbing perpends

Solid brick walls

A large number of bonds is available for one-brick walls and above. They have been discussed in Chapter 8, Level 1. Please refer back to this chapter for further information.

This chapter will deal with the construction of solid walls and, where appropriate, the bonding arrangements.

Basic skills for one-brick walls

Acquiring the many skills required of a bricklayer will take time and patience. What you require is dexterity: the skilful use of the hands together with eye co-ordination.

Watching your tutor or bricklayer carefully on site, and regular practice over time will help you to attain these skills.

By this point in your training you will have begun to create a grasp of the trowel, either left-handed or right-handed use, be able to pick up the mortar with an easy application and spread it on the wall to produce an economical bed of mortar.

Correctly and safely handling bricks to be laid will also need continuous practice, especially when one-brick walls are being built.

The bricks may have to be turned several times in the hand, especially when laying header courses.

The building of solid walls involves some different skills from those required when building half-brick walls.

With one-brick walls the materials are all bricks, but commons could be required for the back of the wall.

English bond

Walls built in English bond are very strong as no straight joints occur in any part of the wall.

A unit of bond, shown hatched, of a one-brick wall in English bond, is illustrated in Figure 6.24.

Flemish bond

Flemish bond is used mainly for decorative purposes as internal joints make this bond slightly weaker than English bond.

A unit of bond, shown hatched, of a one-brick wall in Flemish bond, is illustrated in Figure 6.25,

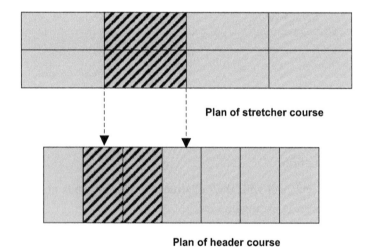

Plan of stretcher course

Plan of header course

FIGURE 6.24
Unit of bond in English bond (no straight joints)

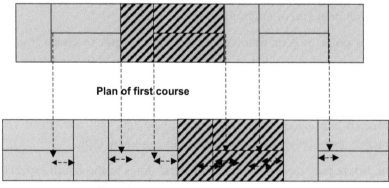

Plan of first course

Plan of second course

FIGURE 6.25
Unit of bond in Flemish bond (straight joints occur on the internal collar joint)

A large number of bonds is available for one-brick walls and above. Some of the most common ones are described below.

Header bond

Header bond is only suitable for walls one brick thick. It consists of all headers, with the bond being formed by three-quarter bats at the quoin, and is generally used in brickwork below ground or walling curved on plan (Figure 6.26).

English bond

Walls built in English bond are very strong as no straight joints occur in any part of the wall.

Alternate courses of headers and stretchers produce quarter-bonds and because of its somewhat monotonous appearance it is used where strength is preferable to appearance.

To achieve and maintain quarter-bond a queen closer must be laid next to the quoin header (Figures 6.27–6.29).

Attention to the following points helps to maintain a good standard of work.

- Cut queen closers neatly and keep them regular in size.

- Keep perpends uniform and plumb as large cross-joints can soon cause you to lose quarter-bond and can bring straight joints on to the face.

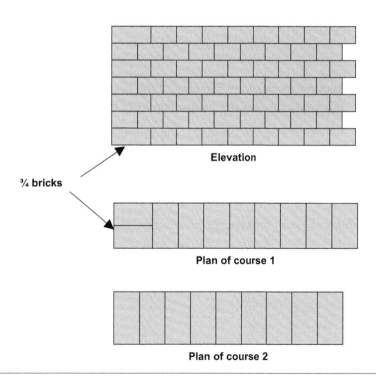

FIGURE 6.26
Header bond

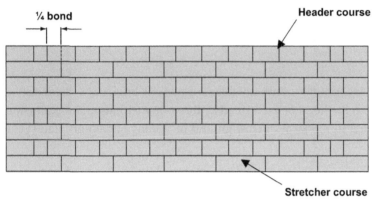

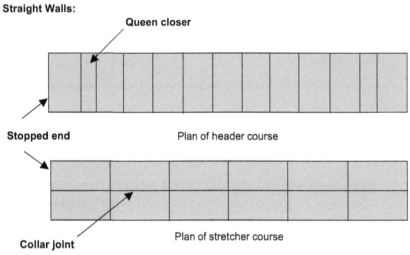

FIGURE 6.27
Bonding arrangements for English bond

- Remove mortar from the back of the bricks against the collar joint as this could prevent the backing up from being laid level.

- When backing up avoid the use of too much mortar near the collar joint, since the backing up could cause the face bricks to move out in front of the line (see Figure 6.21).

Straight walls: Flemish bond

Flemish bond consists of courses of alternate headers and stretchers with the headers in one course placed centrally over the stretcher in the course below (Figures 6.30– 6.32).

A closer is placed next to the quoin header to form the correct quarter-bond.

Straight walls: Dutch bond

Dutch bond is similar to English bond in that it consists of alternate header and stretcher courses, but there are no closers in the header course, and the bond is formed by starting each stretcher course with a three-quarter bat (Figure 6.33).

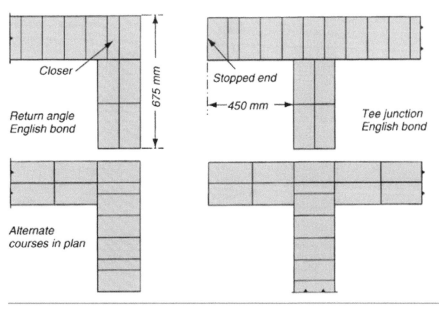

FIGURE 6.28
Return quoins and junctions in English bond

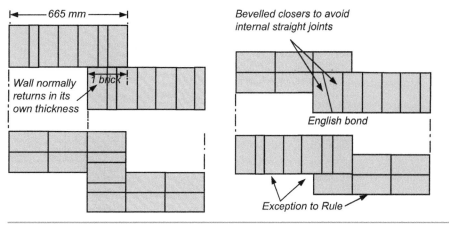

FIGURE 6.29
Double returns in English bond

In addition, the stretcher courses are laid half-bond to each other. This is achieved by placing a header on alternate stretcher courses next to the three-quarter bat.

The perpends in this bond follow each other diagonally across the wall in an unbroken line.

Garden wall bonds
This is a facing bond in which many more stretchers than headers are used. Several internal straight joints occur but a fair face can be achieved on both faces of the wall.

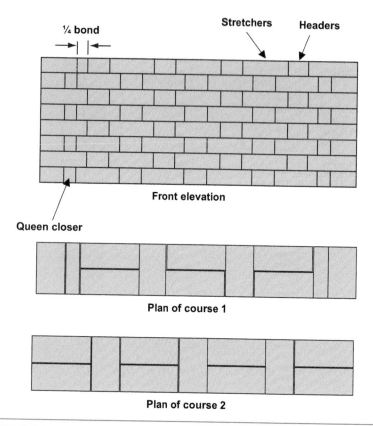

FIGURE 6.30
Bonding arrangements for Flemish bond

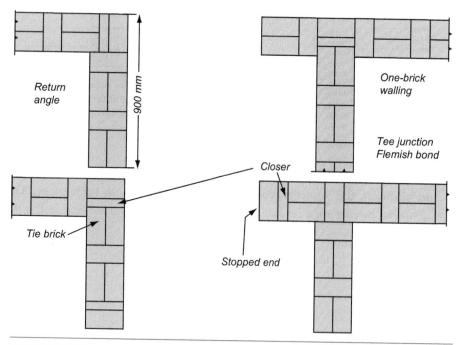

FIGURE 6.31
Return quoins and junctions in Flemish bond

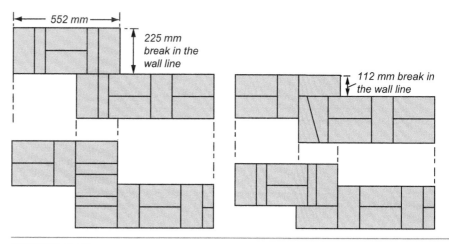

FIGURE 6.32
Double returns in Flemish bond

English garden wall consists of three courses of stretchers to one course of headers. The stretcher courses are laid half-bond with the header course being quarter-bond.

Other versions have up to five courses of stretcher to one course of headers to save bricks.

English garden wall bond is sometimes known as Sussex bond. This is very popular in the USA, where six courses of stretchers are used, and is known there as 6 and 1 bond.

Flemish garden wall consists of three stretchers to one header in each course. To maintain the correct bond, the header in one course must be in the centre of the middle stretcher in the courses above and below.

Flemish garden wall bond is stronger than English garden wall bond as the headers are more evenly distributed.

Although garden wall bonds are occasionally used as face bonds, their main aim is in one-brick-thick walls where a face side of neat brickwork is required on both sides of the wall.

The varying brick lengths make it difficult to keep a fair face on both sides of the wall unless the number of headers is greatly reduced.

If the strength of the wall is of secondary importance, then the number of stretcher courses to each header course in English garden wall bond may be increased to five instead of three.

Similarly, in Flemish garden wall bond, the number of stretchers to one header in each course may be increased to five.

English garden wall bond

English garden wall bond consists of between three and five courses of stretchers to one course of headers.

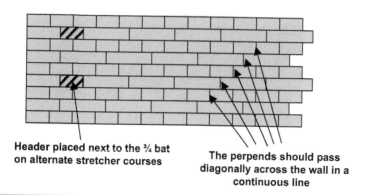

Header placed next to the ¾ bat on alternate stretcher courses

The perpends should pass diagonally across the wall in a continuous line

FIGURE 6.33
Dutch bond

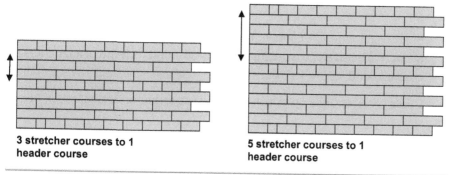

3 stretcher courses to 1 header course

5 stretcher courses to 1 header course

FIGURE 6.34
Variations of English garden wall bond

The stretcher courses are laid half-bond to each other (Figure 6.34).

Flemish garden wall bond

Flemish garden wall bond consists of three or sometimes five stretchers to one header in each course.

The header in one course is laid centrally over the middle stretcher in the course immediately below (Figure 6.35).

Setting out bonds

It is essential that the bricklayer understands fully the reason why walls should be set out before laying any bricks.

The basic principles have been dealt with, but there will be situations when it will not be possible to adhere to the rules, at quoins, junctions and stopped ends.

When a wall is to be built which is not a standard brick length then broken bond or reverse bond may have to be used.

3 stretchers to 1 header in each course

Bonding from the quoin

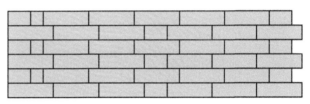

5 stretchers to 1 header in each course

FIGURE 6.35
Variations of Flemish garden wall bond

Broken bond

This can happen in short lengths of brickwork between windows, doors and openings where the walls are not in multiple brick lengths.

Broken bond can sometimes be avoided by 'tightening' or 'opening' the joints. In doing so, bricklayers should work to the brick size plus a 10 mm joint = 225 mm.

It if far better if broken bonds can be avoided altogether, but where openings and piers do not allow for brick sizes, the broken bonds should be carefully set out. Once their position has been decided, it should be maintained throughout the height of the wall.

On no account should a closer be built in the middle of the wall; it should only be placed next to the quoin header.

If a wall length is 56 mm less than brick sizes a half-bat and three-quarters should be used in the centre of the wall.

Remember

The cut brick should not be less than half a brick.

Stretcher bond

When the wall length in stretcher bond is not in brick sizes broken bond can be included, but should be positioned in the centre of the wall.

Three variations of broken bond are shown in Figures 6.36–6.38.

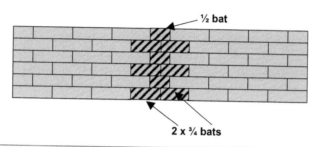

FIGURE 6.36
Broken bond in stretcher bond: two three-quarter-bats in one course and a header in the next

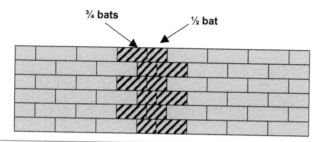

FIGURE 6.37
Broken bond in stretcher bond: a three-quarter-bat and a header in each course

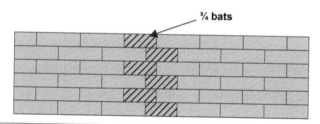

FIGURE 6.38
Broken bond in stretcher bond: a three-quarter-bat in each course

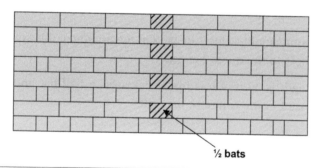

FIGURE 6.39
Broken bond in English bond: half-bats in stretcher courses

English bond

The example in Figure 6.39 shows a wall in English bond which, on the first header course, works out correctly.

But when setting out the alternate courses, working from each corner it leaves a header in the centre of the wall, i.e. broken bond.

A three-quarter bat may be used in both the header and stretcher courses (Figure 6.40).

Flemish bond

When setting out Flemish bonds the stretchers can be shortened to three-quarters to accommodate the broken bond (Figure 6.41). In the next example (Figure 6.42) an extra header is inserted into each course, showing two headers together, to maintain bond.

A header and three-quarter bat have been inserted to maintain bond when the wall dimension is 56 mm less than brick sizes (Figure 6.43).

Two headers and a three-quarter bat have been inserted into one course and a three- quarter bat and an extra stretcher in the alternate course, when the wall dimension is 56 mm more than brick sizes (Figure 6.44).

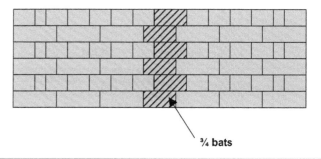

¾ **bats**

FIGURE 6.40
Broken bond in English bond: three-quarter-bats in header and stretcher courses

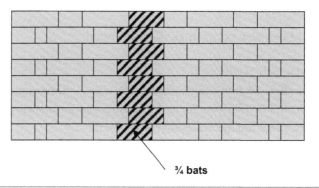

¾ **bats**

FIGURE 6.41
Broken bond in Flemish bond: a three-quarter-bat in each course

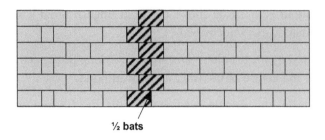

FIGURE 6.42
Broken bond in Flemish bond: an extra header in each course

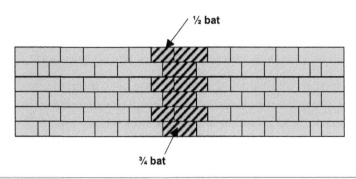

FIGURE 6.43
Broken bond in Flemish bond (wall dimension 56 mm shorter than brick sizes)

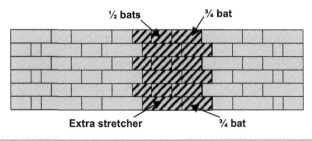

FIGURE 6.44
Broken bond in Flemish bond (wall dimension 56 mm longer than brick sizes)

Garden wall bonds
Broken bond can also occur in garden wall bonds.

English garden wall bond
When the wall is 56 mm longer than brick sizes, a three-quarter bat can be inserted into the header course, and a header and a three-quarter bat can be inserted into the stretcher courses (Figure 6.45).

When the wall is 56 mm shorter than brick sizes, a three-quarter bat can be inserted into the header course instead of a header, and a three-quarter bat can be inserted into the stretcher courses instead of a stretcher (Figure 6.46).

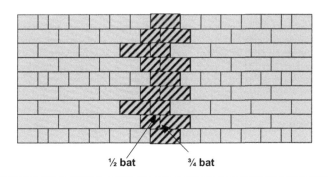

FIGURE 6.45
Broken bond in English garden wall bond (wall dimension 56 mm longer than brick sizes)

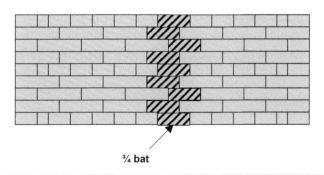

FIGURE 6.46
Broken bond in English garden wall bond (wall dimension 56 mm shorter than brick sizes)

Flemish garden wall bond

When the wall is 56 mm longer than brick sizes, a three-quarter bat and header can be inserted into one course instead of a stretcher, and a three-quarter bat can be inserted into the stretcher courses instead of a header (Figure 6.47).

Another method of using three-quarter bats when the wall is longer than brick sizes is shown in Figure 6.48.

Reverse bond

It is normal in bonding to start both ends of the wall the same way.

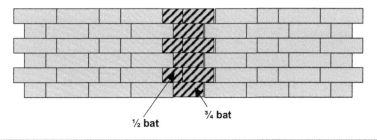

FIGURE 6.47
Broken bond in Flemish garden wall bond (wall dimension 56 mm longer than brick sizes)

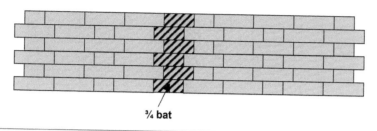

¾ **bat**

FIGURE 6.48
Broken bond in Flemish garden wall bond (wall dimension 56 mm longer than brick sizes)

If you have a stretcher at one end on the first course you should have a stretcher at the other end (Figure 6.49).

On occasions it is possible to change the bonding at the ends of the wall. This would be done to avoid cutting in the centre of the wall. For instance, if you bond the wall dry and you had a half-brick in the centre you would take it to one end.

This is known as reverse bond (Figure 6.50).

English bond

When the dimension of a wall is not multiples of bricks the end bricks can be reversed to avoid cut bricks in the centre of the wall.

The example shown in Figure 6.51 avoids the use of a three-quarter brick in the centre of the wall.

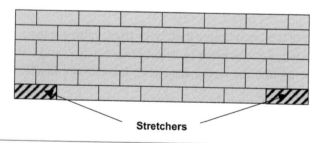

Stretchers

FIGURE 6.49
Normal stretcher bond

Header   Stretcher

FIGURE 6.50
Reverse stretcher bond

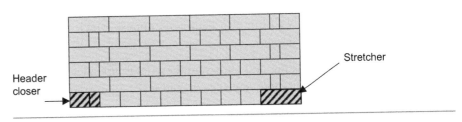

FIGURE 6.51
Reverse English bond

Flemish bond

The example in Figure 6.52 again avoids the use of a three-quarter brick in the centre of the wall.

Boundary walls

These are described as 'free standing' because the brickwork receives no support from floors or roof structure, as do the walls in a building.

For this reason local authority building control department's approval is usually required if a wall higher than 1 m is to be built next to a public road or 2 m high elsewhere, to check that the design is stable. Boundary walls include attached piers, detached piers, gates and raking cuts. Attached piers must be properly bonded to the wall

The DPC for boundary walls is best provided by using two courses of black or red class A engineering bricks, bedded and jointed in cement mortar. The brickwork above and below is solidly bonded to these DPC bricks. The result bonds better than if separated by a flexible DPC, which in a free-standing boundary wall creates a plane of weakness.

Detached piers

Solid free-standing (detached) piers are capable of transferring considerable compressive loads through their structure down to the concrete foundation. Although this type of wall can withstand heavy loads it also has to resist other forces. It has to withstand soil movements, pressure of soil placed behind the wall, inclement weather applied to all faces of the wall and side load pressure.

There must always be adequate foundations provided at the correct depth to resist frost. If the foundations are not deep enough problems with soil movement will be substantial. Cracks will appear and cause irreversible structural damage.

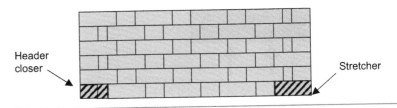

FIGURE 6.52
Reverse Flemish bond

Many piers and boundary walls have soil placed against them which not only applies pressure to the wall but also causes dampness which eventually will cause damage to the brickwork.

The pier has four sides a top and base, and these faces are all subjected to the weather, i.e. rain, wind, snow, etc. The whole wall is constantly wet or dry throughout the year and this can cause considerable damage.

Piers used to support gates, fences, etc., can also have side pressure placed on them. If a side load is placed on the pier it creates a totally different set of stresses which cannot be carried by the brickwork.

Possible problems with piers

Inclement weather can affect the pier on all sides, including the base if a DPC course is not included (Figure 6.53).

A free-standing brick pier or isolated pier is a pillar of brickwork which stands alone. It can be used to carry the ends of beams over a large span or to support garden gates, etc.

Piers can vary considerably in size and bond and can be very difficult to build because of the plumbing points.

Remember

Brickwork is strong in compression but weak in tension.

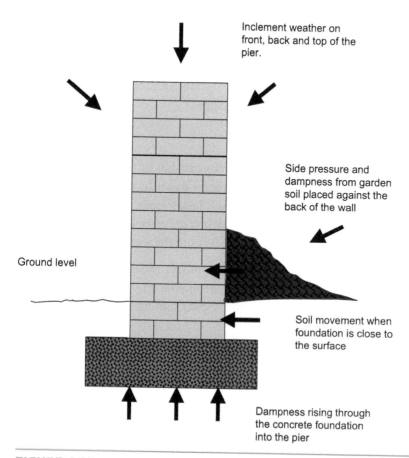

Inclement weather on front, back and top of the pier.

Side pressure and dampness from garden soil placed against the back of the wall

Ground level

Soil movement when foundation is close to the surface

Dampness rising through the concrete foundation into the pier

FIGURE 6.53

Problems with detached piers

The bond specified for a particular pier may have to be amended slightly to avoid the excessive cutting required by some bonds. Excessive cutting is expensive not only on materials but also on time. When there are many cuts the bricklayer has to take time to ensure that all the cuts are the same so as to avoid any irregularity in the perpends or plumbing.

Small-section piers are often the most difficult to build owing to the limited number of perpends.

If several piers are in line, the piers at either end should be built first and then a line attached to either end pier so that the remainder of piers can all be built in line, both front and back.

Always set the piers out square and to brick sizes.

Hollow piers

Piers are often constructed in stretcher bond with a void in the middle (Figure 6.54).

Construction

All piers are difficult to build and require extra skill. Hollow piers are easier than solid piers because of the void in the centre.

Once the setting out has been completed the pier can be built up to DPC level. All detached piers have eight plumbing points, as shown in Figure 6.54.

Each one must be accurate. It is essential that other corners are not pushed out while plumbing. Never leave the plumbing too late; always plumb before the mortar sets. DPCs should always be engineering bricks as these provide better resistance to over- turning of the pier than flexible DPCs.

Overcoming problems

The DPC should be built in engineering bricks. Ensure that soil is not placed at the rear of the wall. Ensure the correct thickness and depth for concrete

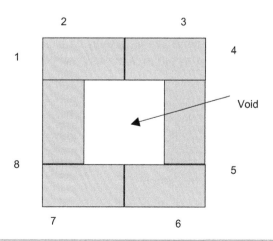

FIGURE 6.54
Two-brick pier in stretcher bond

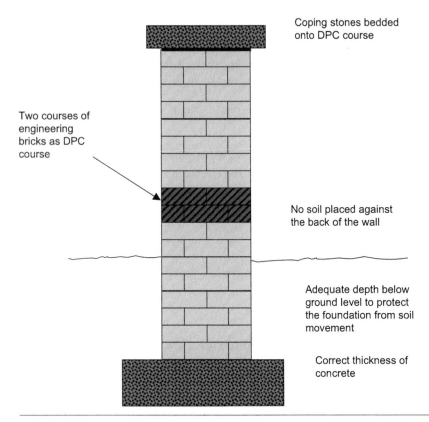

Coping stones bedded onto DPC course

Two courses of engineering bricks as DPC course

No soil placed against the back of the wall

Adequate depth below ground level to protect the foundation from soil movement

Correct thickness of concrete

FIGURE 6.55
Detached pier with problems cured

foundations. The top of wall should be protected by coping stones (Figure 6.55).

Solid English bond piers

Piers built in English bond are solid and usually 1½ bricks square or larger. See the examples in Figure 6.56 showing 1½-brick and 2-brick piers.

Flemish bond piers

Piers built in Flemish bond are also solid and usually 1½ bricks square or larger. See the examples in Figure 6.57 showing 1½ × 1-brick and 2 × 2-brick piers.

Construction of English bond piers

Solid piers are more difficult to build than hollow piers.

There are also eight corners to plumb, but because the pier is more solid there is no room for internal movement.

Always lay the corner bricks first and ensure that they are plumb, level and gauge.

When laying the corner bricks ensure they are immediately gauged and plumb (Figures 6.58 and 6.59).

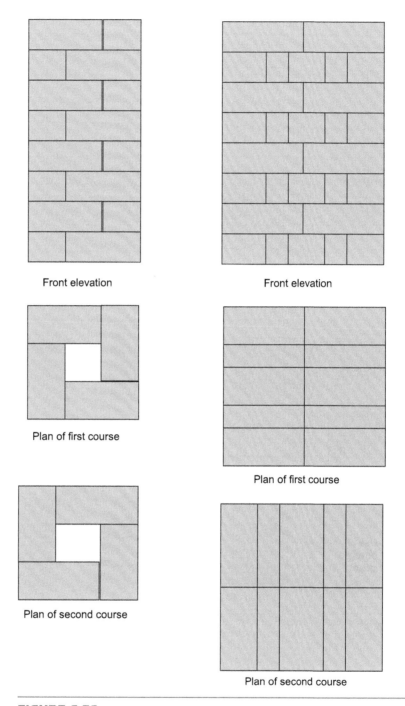

FIGURE 6.56
Examples of bonding to piers

It is essential that the corner bricks are not disturbed while filling in the remainder of the course.

Complete one full course at a time.

When the corner bricks have been laid, plumbed, levelled and gauged the rest of the bricks can be filled in (Figure 6.60). These are laid with caution so as not to disturb the corner bricks.

There is no correct method for filling in the headers and queen closers. You can lay the two headers as shown above, making sure they are levelled and ranged in. Then lay the four queen closers.

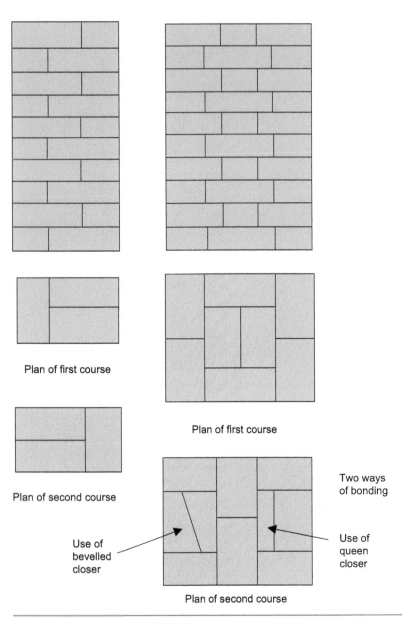

Plan of first course

Plan of first course

Plan of second course

Two ways
of bonding

Use of
bevelled
closer

Use of
queen
closer

Plan of second course

FIGURE 6.57
Examples of bonding to piers

Attached piers

Purely for economic reasons, most boundary walls are constructed in thin walling in stretcher bond and strengthened with attached piers (Figure 6.61).

The attached piers help to prevent the boundary walls from overturning with pressure. Attached piers should not be less than 100 mm thick and must be bonded into the wall (Figure 6.62).

Attached piers could project on one or both sides. They should project by at least half a brick to make the bonding easier.

The DPC should continue through all piers.

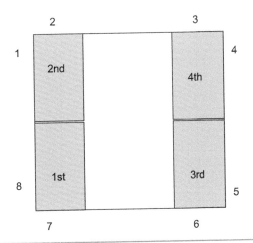

FIGURE 6.58
Building piers

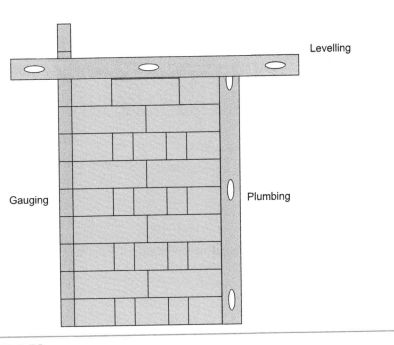

FIGURE 6.59
Checking piers

Construction of boundary walls

Stretcher bond

Attached piers in stretcher bond are the most common type of pier.

The most important consideration with this type of pier is the appearance of the face bond. If maintaining the correct bond on the face of the wall is essential then the example shown in Figure 6.63 would be one solution.

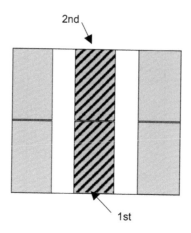

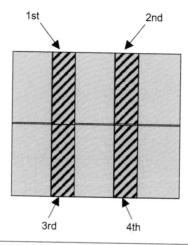

FIGURE 6.60
Building piers

The pier can be strengthened by bedding reinforcement over the pier on alternate courses.

Solid brick walls

Sometimes garden walls can be built one brick wide with two brick attached piers (Figure 6.64).

Attached piers can be used on other brick walls to provide stability, e.g. to long or high walls. The important part is the bonding of the attached piers to the main wall.

The building of the attached piers needs care to ensure plumb and level. The gauge should not need checking as the pier should be built along with the main wall to a line.

As with detached piers, take care not to push the back of the pier when plumbing the front, and vice versa.

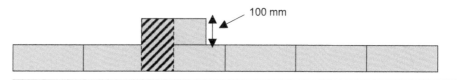

FIGURE 6.61
Bonding single attached piers (dotted bricks show second course)

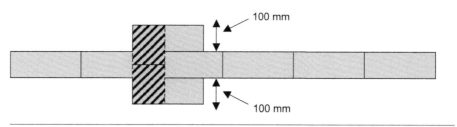

FIGURE 6.62
Bonding double attached piers (dotted bricks show second course)

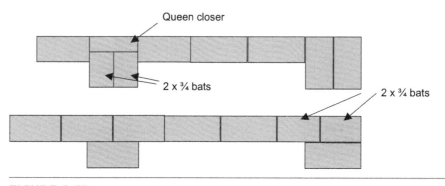

FIGURE 6.63
Bonding attached piers in stretcher bond

Ramps

Ramps are decorative features that could be added to boundary walls, and which the bricklayer must be able to construct.

Ramps are usually next to the pier when gates are being fixed, to allow the pier to be built higher than the boundary wall to support the gates.

Straight ramps can be set out with the use of bricklayer's lines and pins to the required angle (Figure 6.65).

Cutting to rake

A raking cut is required on walls that have to be finished with a sloping end, generally on gable ends, wall ramps and stair strings.

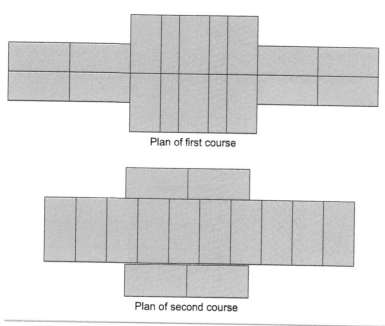

Plan of first course

Plan of second course

FIGURE 6.64
Bonding attached piers in English bond

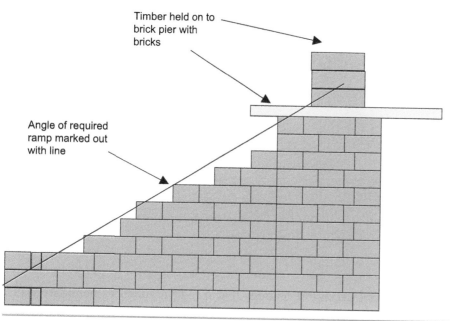

Timber held on to
brick pier with
bricks

Angle of required
ramp marked out
with line

FIGURE 6.65
Setting out cutting to rake

On boundary walls the brickwork is raised to the full height of the raking cut by raking back and toothing the wall, finishing just short of the required line of cut.

Each brick to be cut should be bedded onto the wall in its correct position and then marked in pencil ready to be cut.

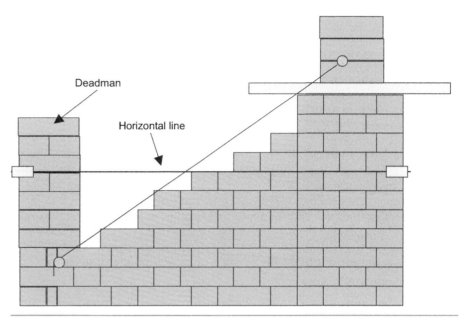

FIGURE 6.66
Use of a deadman to support the horizontal line

Use softing to prevent damage and cut with a hammer and bolster or a mechanical saw.

Deadman

Deadman is the term given to a temporary brick pier bedded to gauge and plumbed to the face of the wall. This acts as a corner and allows a horizontal line to be extended beyond the line of the raking cut (Figure 6.66). A profile could also be used (Figure 6.67).

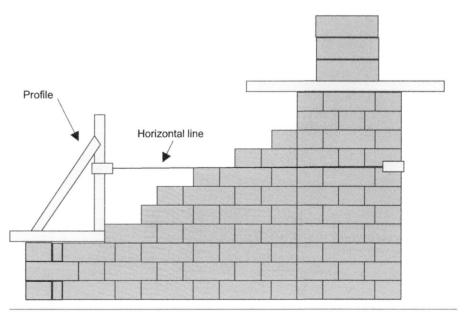

FIGURE 6.67
Use of a profile to support the horizontal line

Procedure for marking and cutting bricks to rake

1. Set the sliding bevel at the necessary angle of the raking cutting.

2. Measure the top edge of the cut brick required.

3. Deduct 10 mm for one cross-joint.

4. Mark in pencil the remaining measurement on the brick to be cut, back and front.

5. Draw in pencil the angle of the sloping (raking) cut, using the sliding bevel, also back and front, taking care to match the slope on the face (Figure 6.68).

6. Cut the brick with hammer and bolster, or a mechanical masonry saw if available.

7. Trim the cut face with a scutch or comb hammer if necessary to avoid any projections.

8. Carefully bed the cut brick to line and level. See examples of marking in Figure 6.69.

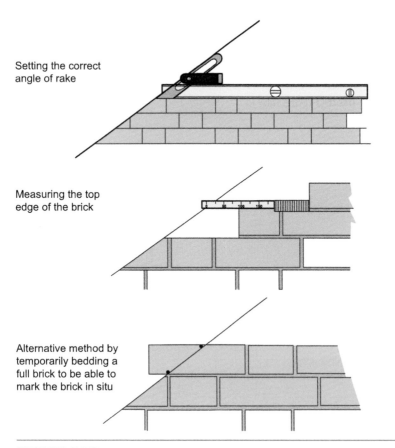

Setting the correct angle of rake

Measuring the top edge of the brick

Alternative method by temporarily bedding a full brick to be able to mark the brick in situ

FIGURE 6.68

Marking the cut bricks

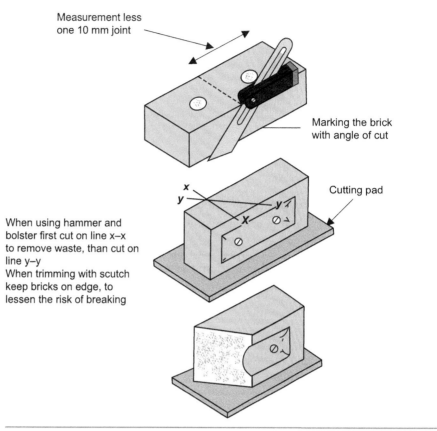

Measurement less one 10 mm joint

Marking the brick with angle of cut

Cutting pad

When using hammer and bolster first cut on line x–x to remove waste, than cut on line y–y
When trimming with scutch keep bricks on edge, to lessen the risk of breaking

FIGURE 6.69
Marking and cutting the bricks

Gate pillars

As mentioned in the previous sections on detached and attached piers, gates and ornamental ironwork are very often built into the piers.

It is essential that the boundary wall is thickened at any point where gates are being fixed so that it is strong enough to withstand the extra pressures applied.

Occasionally the pier supporting a gate will be a detached pier, but more often it is attached to the boundary wall (Figure 6.70).

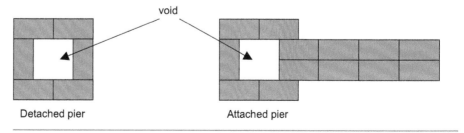

void

Detached pier

Attached pier

FIGURE 6.70
Hollow gate pillars

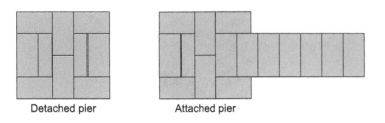

Detached pier   Attached pier

FIGURE 6.71
Solid gate pillars

Because of the stresses placed on the piers by the gates the piers have to be extra strong. One method of achieving this is to construct the piers hollow and infill with concrete.

The concrete can also be reinforced and even attached to the foundation for extra resistance to side movement. Alternatively, the piers may be built of solid brickwork (Figure 6.71).

Hollow piers

Detached and attached piers can be strengthened by the introduction of re-inforced concrete into the void (Figure 6.72).

The brick pier is constructed as a hollow pier, therefore allowing reinforced concrete to be placed into the pier to provide additional strength.

It is important to allow the brick pier to set hard before pouring the concrete, or damage may be done to the pier.

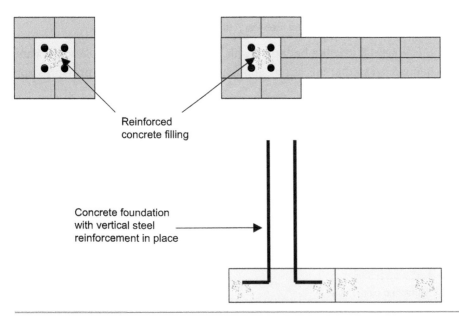

Reinforced
concrete filling

Concrete foundation
with vertical steel
reinforcement in place

FIGURE 6.72
Reinforcing gate pillars

The concrete will require compacting into the pier to achieve maximum strength.

Steel reinforcement should be placed into the concrete foundation when it is being laid. The pier will then have to built around it.

It is essential to ensure that the void in the pier is kept free from mortar droppings, which would affect the finished strength of the pier.

Fixing for gates

When piers are being built to receive gates it is essential to plan beforehand.

Gates can be fixed after the brickwork has hardened or built in as the work proceeds (Figure 6.73).

Fixings for gates can be fitted after the piers have been built by either:

- building in sand courses or
- drilling and fixing hinges.

To fix the brackets after the brickwork has set the sand courses have to be removed and all the sand must be brushed out to leave a clean recess.

The recess should be dampened to ensure adhesion of the mortar.

The gates could be erected in position to identify the correct position of the brackets. These can then be fitted and the brickwork replaced as solidly as possible to ensure that there is no movement when the gate swings.

Fixing as the work proceeds

When gates are being fixed as the work proceeds they can be erected in position and supported in their final position by timbers (Figure 6.74). The

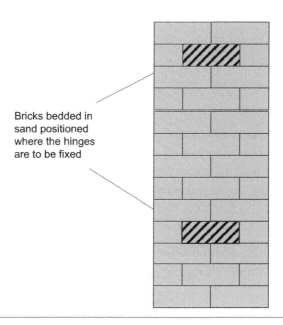

Bricks bedded in sand positioned where the hinges are to be fixed

FIGURE 6.73
Building in sand courses

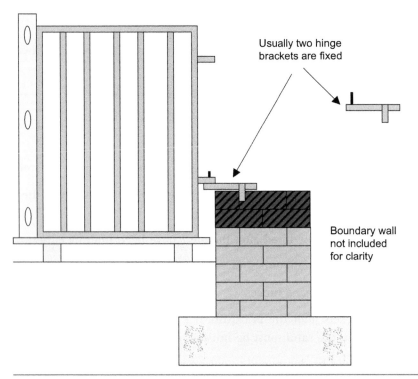

Usually two hinge
brackets are fixed

Boundary wall
not included
for clarity

FIGURE 6.74
Fixing gates to pillars

brackets can then be built in as the work proceeds and could also be allowed to pass into the void so they would be further attached to the concrete infill. The piers must be constructed absolutely plumb, otherwise the gates will not swing correctly. Once the gate has been placed in its final position on timbers it can be plumbed and supported with raking struts.

The brickwork can be built up to the first bracket; this may include the DPC.

The bracket is then built in, allowing it to extend into the void, and could be fixed to the vertical reinforcement.

The brickwork is then built up to the next bracket, which fitted in the same way.

Double gates are fitted in exactly the same way (Figure 6.75). Care again should be taken to set the gates in their required position on battens and support them in an upright position with raking struts. Alternatively, sand courses could be built in and the brackets fixed afterwards.

Copings

Copings are used to protect the tops of walls and piers from moisture penetration; for example, the tops of boundary walls and parapet walls (Figure 6.76).

Materials

Precast concrete copings or standard bricks may be used. The use of engineering bricks is advisable in exposed situations.

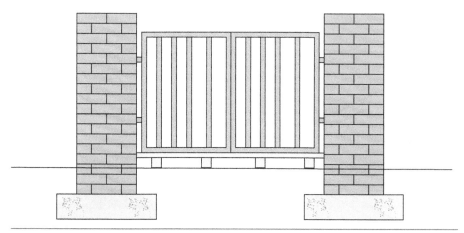

FIGURE 6.75
Fixing gates to pillars

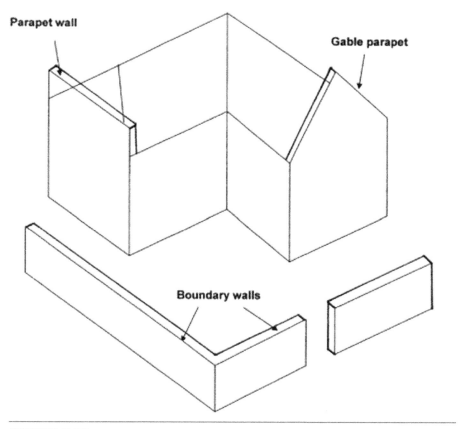

FIGURE 6.76
Positions of copings

Bricks

Standard well-burnt bricks or engineering class bricks may be used as copings.

Precast concrete copings

These are available in standard shapes and lengths or may be specially produced (Figure 6.77). Concrete copings are available in various shapes. Each

(a)   (b)

(c)   (d)

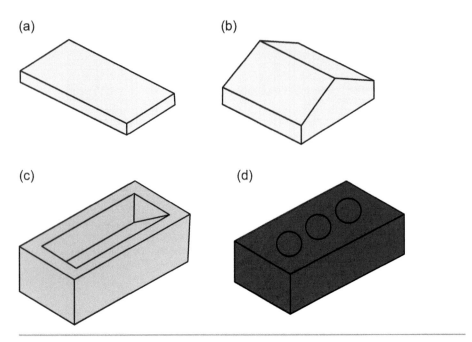

FIGURE 6.77
Coping materials: (a) flat coping; (b) saddleback coping; (c) standard brick; (d) engineering quality brick

provides excellent protection if fitted correctly (Figure 6.78). Flexible DPCs should be provided below the copings.

- Saddleback coping (Figure 6.78a) – The most effective method is to cap the wall with an overhanging concrete coping.

- Feather edge coping (Figure 6.78b) – This needs to be placed so the water runs off onto your own property. A good overhang is essential to run the water clear of the brickwork.

- Detail (Figure 6.78c) – The drip, a groove along the underside of the coping, prevents the passage of water from the top of the coping, the weathering, to the face of the wall.

Lime staining

Lime staining is caused when free lime leaches from mortar or concrete products under wet conditions (Figure 6.79). The lime accumulates on the face of the brickwork, leaving unsightly white patches along the path of the water that carries it.

Lime staining is very often found below the cross-joints of concrete coping stones as a result of the drip being blocked.

When cross-joints are applied take special care not to block the drip groove as this will cause moisture to run back onto the face of the wall, where lime staining will occur.

Positioning copings

Bed and position the end copings and ensure that they are level.

> **Remember**
>
> Flexible DPCs should be sandwiched between mortar beds.

> **Remember**
>
> Lime leaching will not wear off. has to be removed by cleani with a diluted solution of hydr chloric acid.
>
> Always wear the correct protecti clothing and equipment wh cleaning brickwork with acids.

(a)

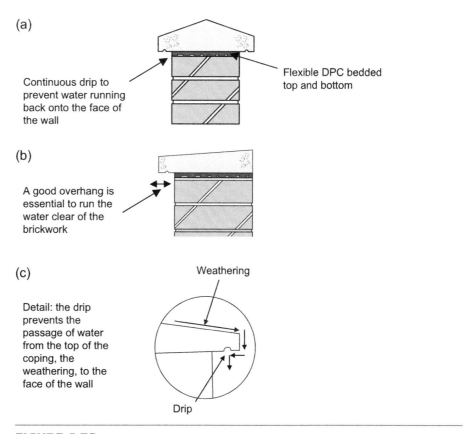

Continuous drip to prevent water running back onto the face of the wall

Flexible DPC bedded top and bottom

(b)

A good overhang is essential to run the water clear of the brickwork

(c)

Detail: the drip prevents the passage of water from the top of the coping, the weathering, to the face of the wall

Weathering

Drip

FIGURE 6.78
Various coping details: (a) overhanging concrete coping; (b) feather edge coping; (c) detail

Attach building lines to front and back edge if wide copings are being laid. For narrow copings a front line is sufficient.

The line is positioned to the eyeline, which is either the upper or lower edge of the coping, depending on its height in relation to where it is viewed from.

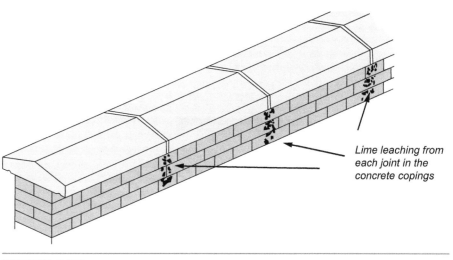

Lime leaching from each joint in the concrete copings

FIGURE 6.79
Lime leaching

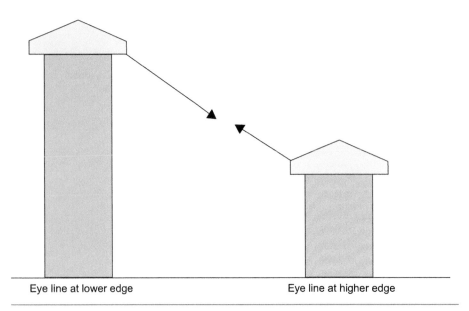

FIGURE 6.80
Eyeline of copings

The eyeline is determined by which edge is seen by the eye. As a general rule, copings above head height have the eyeline to the lower edge. See the example in Figure 6.80.

To lay copings, spread a bed joint and butter the joint on the previously laid coping.

Position the coping, making any necessary adjustments by light tapping with a rubber- headed hammer or onto a piece of timber to protect the coping from damage.

Brick capping

Protection can be provided using a brick on edge and tile creasing. These are not as effective as concrete copings owing to the increase in cross-joints.

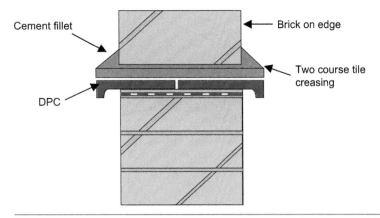

FIGURE 6.81
Brick on edge with tile creasing

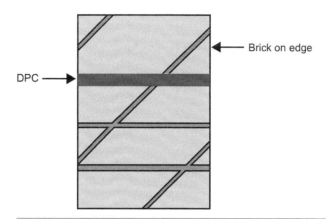

FIGURE 6.82
Brick on edge

Bricks should be class A or B engineering quality, bedded in a strong cement–sand mix.

Walls can be capped with a brick on edge without an oversailing course. This is not a very satisfactory method as it allows the water to penetrate the top surface and then run down the face of the wall (Figures 6.81 and 6.82).

Construction

When using a brick on edge, with or without tile creasing, the flexible DPC should be correctly bedded onto the boundary wall.

The tile used is usually the nibless type, but with care the bottom course could have nibs to form a decorative effect.

Lay the first course to line with the nibs projecting downwards. This course should be lined in along the lower edge.

The second course should have the nibs removed and be laid half-bond and lined in along the upper edge.

Prepare three bricks at each end of the wall, unless special stop ends are being used (Figure 6.83).

The remainder of the bricks should be laid to line, taking care to check the gauge (Figure 6.84), and ensuring that the vertical joints are fully buttered and any frogs filled.

> **Note**
>
> When engineering class bricks are used as a brick on edge, the flexible DPC may be omitted.

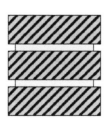 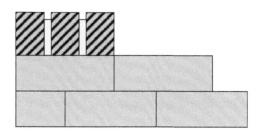

FIGURE 6.83
End block of three bricks bedded to gauge on floor

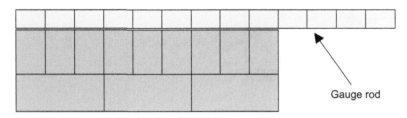

FIGURE 6.84
Bricks laid to horizontal gauge

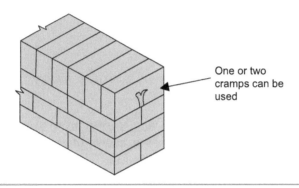

FIGURE 6.85
Protection to end bricks

A regular joint size must be maintained on the brick-on-edge course and no cuts are permitted.

If the length of a wall to receive the brick on edge is composed of whole stretchers in any course, then each stretcher will accommodate three bricks on edge together with their joints.

At stopped ends cramps or retainers can be used and are bedded in below the brick-on-edge course (Figure 6.85). These help to protect the vulnerable corner bricks.

Decorative features

A bricklayer will have to construct many decorative features during work activities; for example:

- string courses
- oversailing courses.

String courses

These are horizontal courses built into the face of a wall to form a decorative feature.

A soldier course is a form of string course, when it continues around a building.

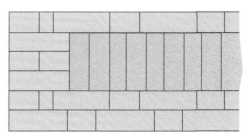

Elevation of soldier string course

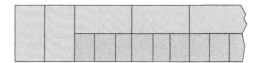

Plan of soldier string course

FIGURE 6.86
Soldier string course

An example is shown in Figure 6.86.

Construction

Because soldier courses are classed as decorative features, good selection of the bricks is essential.

Ensure that they are all the same length, accuracy of shape, colour and texture. Reject any bricks with bows, twists, chips, etc.

The course the bricks sit on should be gauged and marked out accordingly to receive the appropriate number of soldiers. A pencil should be used to mark the gauge on the bricks.

All the bricks should have buttered, slightly furrowed joints.

Take care not to smudge the face of the bricks.

Tap the brick into position and check immediately for plumb with a small boat

> **Remember**
>
> If the opening does not work then gauge; tighten the joints rather than open them.

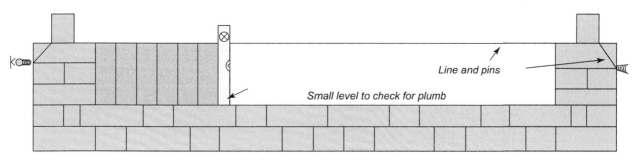

Craft operation for soldier course

FIGURE 6.87
Constructing a soldier string course

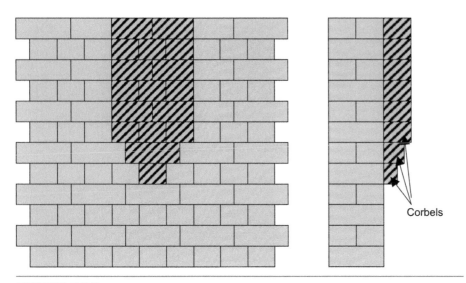

FIGURE 6.88

A simple form of corbel to form a decorative feature

level. Because most soldier courses are built at eye level it is essential to plumb every brick, otherwise a very poor effect will be produced and the bricks will appear to be leaning.

Lay the bricks from the ends to the centre either side to ensure even weight distribution. Take care when placing the last brick, ensure that it is fully buttered on both sides and gently slide it into position, checking that there is no build-up of mortar on the head of the opening (Figure 6.87).

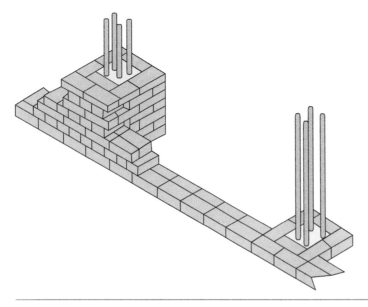

FIGURE 6.89

Vertical reinforcement

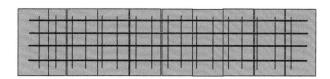

FIGURE 6.90
Horizontal reinforcement

Oversailing courses

Oversailing courses or corbels is the term given to a brick or several bricks which project from the face of a wall (Figure 6.88).

Oversailing or corbelling work may be used to thicken a wall or form a decorative feature.

When this type of work is undertaken it is important that the underside arris of each course is level because it is this line that is seen by the eye.

It is good practice to use headers for corbelling out and restrict the projection to a quarter of a brick.

The Building Regulations state that the extent to which any part of a wall overhangs the wall below shall not be such as to impair the stability of the wall or any part of it.

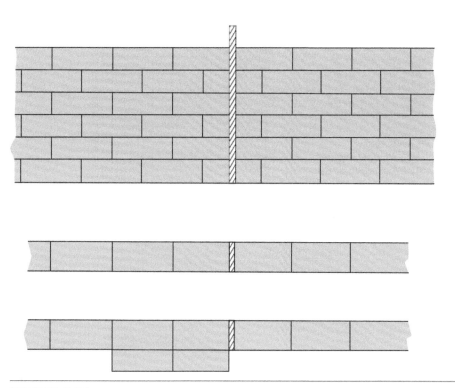

FIGURE 6.91
Vertical movement joints

To prevent this happening, under no circumstances should a wall overhang more than the thickness of the wall below the corbel.

Reinforcement

Brickwork under normal loading conditions is usually strong enough without reinforcement.

When brickwork is put under pressure which is trying to crush the wall, re-inforcement can be added to withstand these pressures and increase the strength of the brickwork.

Reinforcement can be placed either vertically or horizontally.

We have already dealt with vertical reinforcement to gate pillars in Figure 6.72. Figure 6.89 shows pillars to a boundary wall being reinforced in a similar manner. Vertical reinforcing rods are placed into the foundation concrete and extend the height of the hollow pillars. The hollow pillar is then filled with concrete to produce reinforced brick pillars.

Brick walls can be reinforced by building in horizontal mesh as the work proceeds to strengthen the wall from side pressure (Figure 6.90).

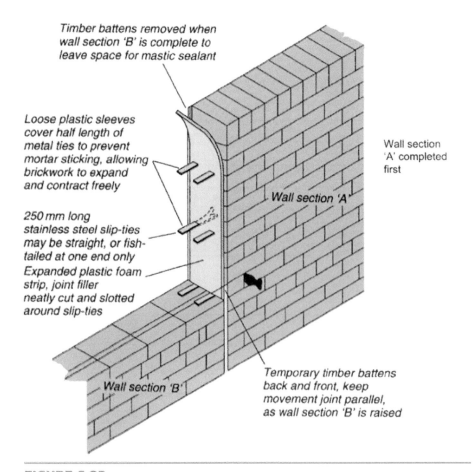

Timber battens removed when wall section 'B' is complete to leave space for mastic sealant

Loose plastic sleeves cover half length of metal ties to prevent mortar sticking, allowing brickwork to expand and contract freely

250 mm long stainless steel slip-ties may be straight, or fish-tailed at one end only

Expanded plastic foam strip, joint filler neatly cut and slotted around slip-ties

Wall section 'A' completed first

Wall section 'A'

Wall section 'B'

Temporary timber battens back and front, keep movement joint parallel, as wall section 'B' is raised

FIGURE 6.92
Building in a vertical movement joint

The wire mesh is bedded into the mortar every third or fifth course as required.

Vertical movement joints

The outer leaf of face brickwork in a wall will expand and contract owing to changes in temperature and moisture content.

Allowance must also be made for sideways expansion and contraction, however, so that the outer leaf of brickwork does not cause damage to itself or the structure.

A typical vertical movement joint, with expanded plastic foam joint filler projecting, can be seen in Figure 6.91. This shows a 102 mm thick brick wall.

Vertical movement joints should be provided at intervals between 9 and 12 m for clay brickwork; not exceeding 7–9 m for sand lime (calcium silicate) bricks; and at 6 m intervals for concrete blockwork.

Clearly, vertical movement joints form a straight joint weakness in a wall, and so must be strengthened with stainless steel slip-ties every fourth course (Figure 6.92).

The expanded plastic foam filler is kept back 12 mm from both faces so that a sealant mastic can be applied to seal these movement joints against rain penetration.

Multiple-choice questions

Self-assessment

This section of the book is designed to allow you to check your level of knowledge. The section consists of revision questions for this chapter. The questions are all multiple choice and have four possible answers. The answers are to be found at the end of the book.

The main type of multiple-choice question will be the four-option multiple-choice question. This will consist of a question or statement, known as the stem, followed by a choice of four different answers, called the responses. Only one of these responses is the correct answer; the others are incorrect and are known as distracters.

You should attempt to answer the questions by choosing either (a), (b), (c) or (d).

Example

The person employed by the local authority to ensure that the Building Regulations are observed is called the:

    (a) clerk of works

    (b) building control officer

    (c) council inspector

    (d) safety officer

The correct answer is the building control officer, and therefore (b) would be the correct response.

Solid walls

Question 1 Identify the following, which could be used to finish off the top of a boundary wall:

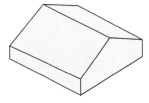

    (a) feather edge coping

    (b) saddle back coping

    (c) gabled coping

    (d) pitched coping

Question 2 Which of the following is the correct definition of English garden wall bond?

    (a) alternate courses of headers and stretchers

    (b) alternate headers and stretchers on each course

    (c) three stretchers to one header in each course

    (d) three courses of stretchers to one course of headers.

Question 3 Identify the following bond:

(a) English bond

(b) Flemish bond

(c) English garden wall bond

(d) Flemish garden wall bond

Question 4 What is a tingle plate used for?

(a) to fasten the lines to the stopped end

(b) to keep the wall plumb

(c) to support the line in the centre of a long wall

(d) to keep the wall level

Question 5 Why is it necessary to load out bricks for a wall from several different packs?

(a) to prevent banding on the walls

(b) to select the best bricks for the wall

(c) to use the bricks as quickly as possible

(d) to ensure the correct number of bricks

Question 6 Identify the following bond:

(a) English bond

(b) Flemish bond

(c) English garden wall bond

(d) Flemish garden wall bond

Question 7 When a cut brick has to be placed in the centre of a wall it is known as:

(a) broken bond

(b) reverse bond

(c) Dutch bond

(d) garden wall bond

Question 8 What is the purpose of the cramp built into the end of a brick on edge course as shown in the drawing?

(a) for decorative purposes

(b) to keep the end of the wall plumb

(c) to protect the end of the wall

(d) to keep the brick on edge level

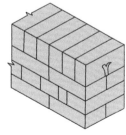

CHAPTER 7

Constructing Cavity Walls with Openings

This chapter will cover the following City and Guilds units:

- Apprenticeship   Module 5   S2.1, S2.2, K2.1, K2.2, K2.3
                    Module 12   S1.1, S2.2, S2.3, S2.4
- Certificate   Unit 202   1.1, 1.2, 1.3, 2.1, 2.2, 2.3, 3.1, 3.2, 3.3, 1.1, 3.2, 4.3, 1.4
- Diploma   Unit 206   1.5, 1.6, 1.7, 1.8, 1.9, 1.10, 1.11
                                      2.1, 2.2, 2.3, 2.4, 2.5, 2.6, 2.7, 2.8, 2.9, 2.10
                                      3.3, 3.4, 3.5 ,3.6, 3.7, 3.8, 3.9, 3.10, 3.11, 3.12
                                      4.1, 4.2, 4.3, 4.5, 4.4, 4.5, 4.6, 4.7, 4.8, 4.9, 4.10, 4.11, 4.12, 4.13
                                      5.1, 5.2, 5.3, 5.4, 5.5, 5.6, 5.7, 5.8, 5.9

This chapter is about:

- Interpreting instructions
- Adopting safe and healthy working practices
- Laying bricks and blocks to line to form cavity walling

The following Skills and Knowledge requirements will be covered:

- Setting out and building cavity walls
- Materials used to build cavity walls
- Methods used to insulate cavity walls
- Methods used to form openings in cavity walls

The following Certificate learning outcomes will be covered:

- Plan prepare and build single leaf walls and cavity walls
- Materials used to build cavity walls
- Methods used to insulate cavity walls
- Methods used to form openings in cavity walls

The following Diploma learning outcomes will be covered:

- Know how to select the required quantity of resources to construct cavity walls
- Know how to construct cavity walls straight and return corners
- Know how to insulate cavity walls
- Know how to form openings in cavity walls

Building information

Definition of cavity walls

The standard form of construction for the external walls of brick buildings is called cavity walling. This means that the bricklayer builds the two separate 'leaves' or 'skins' of brick masonry (a general term indicating brickwork and or blockwork), with a 50–75 mm wide space between.

The outer skin is usually 102.5 mm thick face brickwork, but may be constructed from facing quality blocks. The inner skin is usually 100 mm thick common blocks that are later plastered to receive internal decoration (Figure 7.1).

Both skins of brick masonry are joined together with a regular pattern of corrosion- resistant ties, so that they behave as one single wall.

A continuous space termed a cavity, which is approximately 50 mm wide, is produced and the walls are supported by specially prepared ties of metal called wall ties.

The prevention of the passage of moisture from the outer to the inner walls is of the utmost importance in cavity wall construction, and to achieve this, damp-preventing materials must be inserted in the structure at special points, such as window and door reveals, and window and door heads.

Purpose of cavity walls

Cavity walls are very useful for damp and exposed positions to ensure a dry interior to the building. The reasons for this are:

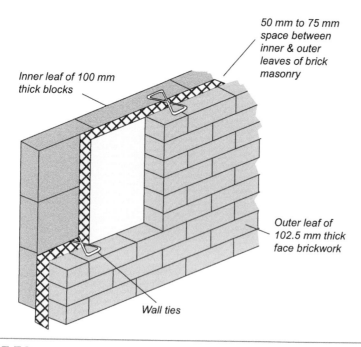

50 mm to 75 mm space between inner & outer leaves of brick masonry

Inner leaf of 100 mm thick blocks

Outer leaf of 102.5 mm thick face brickwork

Wall ties

FIGURE 7.1
Typical insulated cavity wall

- The cavity provides a break between the outer wall of the building, which may damp, and the inner wall.

- To achieve a better balanced temperature inside the building. The cavity keeps the inside cool in the summer and helps to retain heat in the winter.

Where proper care is exercised in the building of this type of wall, it will prevent the passage of moisture from the outer or protective wall to the inner wall.

Cavity wall construction began to be widely used from the 1920s, as a way of preventing dampness from soaking through the outer walls of buildings.

The 50–75 mm wide gap stopped rain penetrating from the outer surface to the plastered inner surface of external walls by capillary action (when water is drawn through hairline channels within a porous structure, e.g. a brick, by the action of surface tension). This was possible when outer walls were commonly 215 mm thick solid brickwork.

As a secondary advantage, this space between the inner and outer skins also provides thermal insulation for modern buildings, because heat energy mainly escapes by conduction through solid material. Air is a poor conductor of heat energy, therefore the rate of heat loss is very much slower than was the case when buildings had solid outer walls (Figure 7.2).

Building Regulations

The basic advantages of cavity wall construction, over solid 215 mm thick brickwork, for the outer walls of a building have been incorporated in the current Building Regulations.

In order to satisfy requirements of the current Building Regulations and enable planning permission to be obtained, cavity wall construction is usually specified whether the structure is low rise or multistorey.

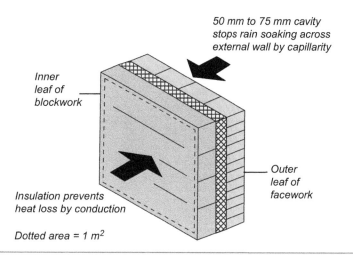

FIGURE 7.2
Function of typical cavity walling

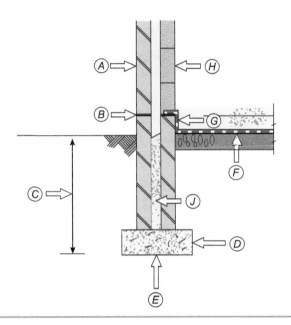

FIGURE 7.3
Cavity walling on standard strip foundation.

Figure 7.3 indicates how the basic requirements of the Building Regulations are satisfied, where standard strip foundations are specified with a solid ground floor slab. Figure 7.4 shows a trench-fill foundation, associated with a suspended ground floor construction of prestressed, precast concrete floor beams supporting standard size concrete blocks.

The following labels are used in Figures 7.3 and 7.4:

A: external cavity wall, providing resistance to through-penetration of rain

B: horizontal damp-proof course (DPC) in both leaves, not less than 150 mm above ground level, to prevent dampness rising from the soil

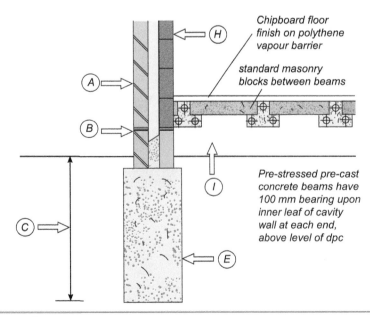

FIGURE 7.4
Cavity walling on trench fill foundation with suspended concrete floor

C: a minimum distance of 1 m between ground level and the underside of the concrete foundation, as a protection against frost heave in winter and drying shrinkage of clay subsoils in summer

D: a minimum 150 mm thickness of foundation concrete to transfer the building load adequately on to the natural foundation of the subsoil

E: sulphate-resisting cement in the foundation concrete and substructure brickwork, necessary where soluble sulphates are present in subsoil water

F: continuous damp-proof membrane (DPM) across the ground floor areas to prevent dampness rising from the soil; ground floor U-value of 0.13; insulation can be placed underneath, on top or as a sandwich layer

G: DPM and DPC, lapped and joined within the floor thickness around the perimeter of rooms

H: external walls built of lightweight blocks and other thermal-insulating material to give a U value of 0.18 (in other words, heat energy must not escape through the outer walls at a rate greater than 0.18 W/m^2/hour, per degree difference in temperature internally and externally)

I: ventilated air space separating the suspended floor from damp soil

J: weak cavity fill to prevent collapse of walls owing to pressure from the ground.

NOTE

Please check with the current Building Regulations for the latest amendments to U-values.

Function of cavity walls

Figure 7.5 shows that it is the inner leaf of cavity walling that largely supports the load from floors and roof in a low-rise building of load-bearing wall

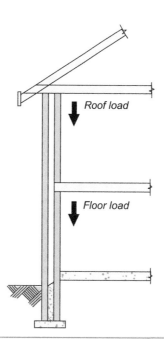

FIGURE 7.5
Loads on cavity walls

construction. Common building blocks have totally replaced common bricks for this inner leaf, owing to improved thermal insulation values and bricklayer output (one standard size 100 mm thick block is equal to six bricks).

The purpose of the outer skin of facework is to give the building a weather-resistant and pleasant appearance, by selection from the wide range of colours and surface textures of bricks and facing blocks available to the designer.

Although cavity walling is constructed with properly finished, solidly filled mortar joints, it is expected that this outer skin will let rain water soak through as far as the cavity. This is because mortar, bricks and blocks are porous to varying degrees. This through-penetration will be highest on those elevations of a building exposed to prevailing wet winds (Figure 7.6).

In the early days of cavity walling it was considered good practice to ventilate the cavity as the flow of air around the cavity would keep it dry. During the 1960s, this theory changed with the increase in fuel prices. There was more emphasis on retaining heat in buildings and the cavities were then sealed to increase the resistance to heat loss through the cavity wall.

This has now changed again as the current Building Regulations are asking for even less passage of heat through the cavities. To achieve this reduction in heat loss cavities are now either partially or totally filled with insulation.

Constructional requirements

Bricks or blocks should be properly bonded and solidly built with mortar not weaker than 1:2:9.

The leaves should be securely tied together with wall ties.

The cavity should not be less than 50 mm or more 75 mm in width at any level.

The leaves should not be less than 100 mm thick at any level.

The overall thickness should not be less than 190 mm.

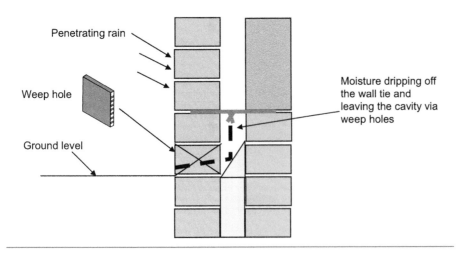

FIGURE 7.6
Use of drip on wall tie

The wall should not exceed 3.5 m in height and 12 m in length, or 9 m in height and 9 m in length.

The inner leaf cannot be less than 75 mm thick if the walls form part of a private dwelling house.

The current Building Regulations specify that the external wall must have a DPC at a height of not less than 150 mm above the finished surface of the adjoining ground.

Advantages of cavity wall construction

- Able to withstand driving rain.

- Gives good thermal insulation.

- Gives good sound insulation.

Disadvantages of cavity wall construction

- Requires a high standard of design and workmanship and supervision.

- Vertical DPC at all openings.

- More expensive than solid walls.

Adopting safe and healthy working practices

Safety legislation has already been covered in Chapter 2. Please refer back to this chapter if necessary.

Particular areas to consider when building cavity walls are:

- personal protective equipment (PPE)

- manual handling of materials

- mechanical tools and equipment

- working at heights.

Personal protective equipment

Depending on the type of workshop or site situation, the wearing of correct safety clothing and safe working practices are the best methods of avoiding accidents or injury. On some sites certain PPE is compulsory.

All construction operatives have a responsibility to safeguard themselves and others. Protecting oneself often means wearing the correct protective clothing and safety equipment.

Manual handling of materials

More than a quarter of all reported accidents are associated with manual handling. Manual handling is defined as the transporting or supporting of loads by hand or bodily force such as human effort.

The Manual Handling Operations Regulations outline how to deal with risks to the safety and health of construction site operatives.

The site operative should be able to select and use appropriate safety equipment and protective clothing when handling different materials.

If the item to be moved is an awkward size or shape the site operative should be able to select and use appropriate equipment or aids to carry materials.

They should also be able to demonstrate safe manual handling techniques.

Movement of materials

Materials can be moved around the site either by hand or by machine.

Most materials are delivered to site on pallets and possibly shrink wrapped. These can then be stored on site, possibly in the compound, until required. They can be moved around the site with a fork-lift truck (Figure 7.7).

Although mechanical lifting methods are becoming more common on larger sites, on small sites materials may still be moved with wheelbarrows.

Working at heights

Working platforms have already been explained in Chapter 2.

The Work at Height Regulations 2005 apply to all work at height where there is a risk of a fall liable to cause personal injury.

A scaffold is a temporary staging to assist bricklayers and other tradespeople to construct a building.

The scaffold must be spacious and strong enough to support people and materials during construction.

FIGURE 7.7
Equipment for moving materials

Many accidents are due to simple faults such as misuse of tools, untied ladders, a missing toeboard, etc.

Unfortunately, a number of accidents are also due to ignoring the Construction (Health, Safety and Welfare) Regulations, which specify basic scaffolding requirements.

The three basic requirements for scaffolds are:

- They should be suitable for the purpose.
- They should be safe.
- They should comply with the regulations.

General safety

No scaffold should be erected or be substantially added to or altered or be dismantled except under the immediate supervision of a competent person and, so far as possible, by competent workers possessing adequate experience of such work.

When work cannot be safely carried out from ground level or from part of a building or other permanent structure, there shall be provided either scaffolds or where appropriate ladders or other means of support, all of which shall be suitable for the purpose.

Access

It is important that you are familiar with a wide range of access equipment.

It is usual to access a bricklayer's scaffold from a ladder, which must be positioned so that this can be done easily and safely.

In the case of access to a scaffold, the ladder should be securely attached to the scaffold, preferably inside the structure (Figure 7.8).

Remember

The Work at Heights Regulations 2005 require that ladders should only be considered where a risk assessment has shown that the use of other more suitable work equipment is not appropriate because of the low risk, and the short duration of the task or considerations of where the work is located.

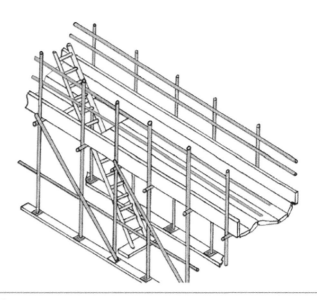

FIGURE 7.8
Access to working platform

Any surface on which the ladder rests should be stable and of sufficient strength to support the ladder so that its rungs remain horizontal and it can support any loading placed on it.

The final rung of the ladder from which the operative steps onto the platform should ideally be just above the surface of the platform.

The ladder, which should be secured at both the top and bottom, should extend at least 1.05 m (approximately five rungs) above the platform. Bricklayers require a scaffold wide enough to allow:

- the stacking of materials
- working space for the bricklayer
- room for the passage of people and materials (Figure 7.9).

Guardrails and toeboards

Access platforms more than 2 m high must have guardrails and toeboards. The top guardrail should be at least 950 mm above the platform with intermediate guardrails fixed so that any gap between the rail and any other means of protection is not greater than 470 mm.

The risk of falling materials causing injury should be minimized by keeping platforms clear of loose materials.

In addition, materials or other objects must be prevented from rolling, or being kicked, off the edges of working platforms. This can be achieved by fi toeboards, solid barriers, brick guards or similar at open edges.

At the end of the working day it is important to turn back the batten nearest the new brick wall to prevent any splashing if it rains (Figure 7.10).

Note

There is no need for a centre guardrail when brickguards are being used

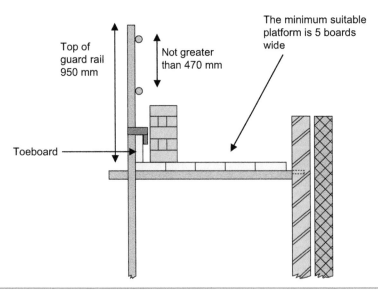

FIGURE 7.9
Bricklayer's working platform

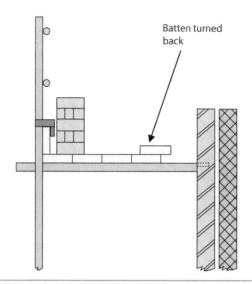

FIGURE 7.10
Inner batten turned back at the end of the working day

Selecting materials, components and equipment

Cavity walls are constructed using bricks and blocks and other components as described in the following paragraphs.

Bricks and blocks

Normal construction consists of facing bricks on the outside wall and lightweight blocks on the inside (Figure 7.11). Variations of this are found with blocks on both walls.

Older versions of cavity walls can be found with common bricks on inside walls.

In response to the requirements of approved Document L, many manufacturers of blocks have produced blocks that help to meet the thermal requirement (Figure 7.12).

Future developments will see the U value fall even more and could result in manufacturers again revisiting the design of blocks.

Blocks for inner walls of cavity walls are available in widths greater than 100 mm, to give greater insulation value

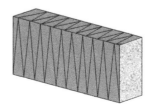

FIGURE 7.11
Bricks and blocks

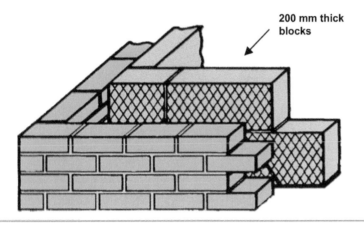

FIGURE 7.12
Typical cavity wall quoin

Wall ties

Wall ties are used to give stability to cavity walls by tying the inner and outer leaves together.

A range of proprietary ties has been produced for this job of tying together inner and outer skins of brick masonry, as 'substitute headers' so that both leaves behave as one wall. These wall ties are made from stainless steel, galvanized steel or polypropylene, so they do not provide a passage for moisture.

Ties are shaped so as to form a drip at their centre and the ends should be designed to be able to be bedded securely in both leaves. Examples are shown in Figure 7.13.

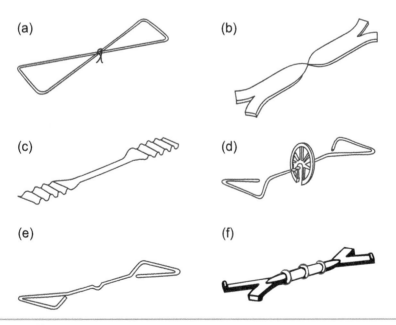

FIGURE 7.13
Types of wall tie: (a) butterfly; (b) galvanized mild steel wall tie; (c) stainless steel tie; (d) double triangle with plastic clips; (e) double triangle; (f) plastic

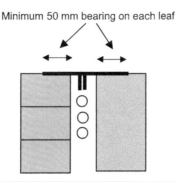

FIGURE 7.14
Position of wall ties

Wall ties should be bedded level or slope slightly towards the outer leaf with the drip positioned downwards in the centre of the cavity (Figure 7.14).

The standard maximum spacing for wall ties is at intervals of 900 mm horizontally and every sixth course vertically.

Each horizontal layer should be offset. For the purposes of estimating quantities of wall ties required, this works out at approximately 2.5 per square metre.

Air bricks

Ventilation into cavities or through cavities is maintained by building air bricks into the external wall. These are available in normal brick sizes and colours to match the facing bricks (Figure 7.15).

Air bricks are built into the external walls along with air ducts. These are built into the external walls at approximately 2 m centres (Figure 7.16).

Thermal insulation

Cavities can be partially filled or total filled with various types of insulation. The insulation is made from various materials and available in various thicknesses such as expanded polystyrene bead board and extruded expanded polystyrene.

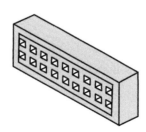

FIGURE 7.15
Air brick

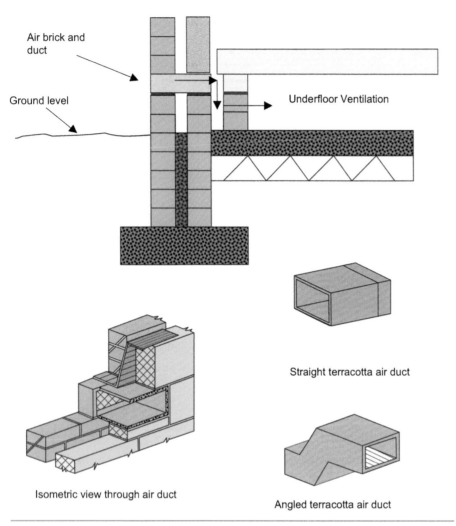

Air brick and duct

Ground level

Underfloor Ventilation

Straight terracotta air duct

Isometric view through air duct

Angled terracotta air duct

FIGURE 7.16
Air brick details

Insulation slabs are usually 450 mm high and 1200 mm long.

For partially filled cavities solid insulation is required so that it can be fixed back to the internal wall with special wall ties (Figure 7.17).

When cavities are totally filled solid or flexible insulation can be used (Figure 7.18).

Door and window frames

Door and window frames are required to be built into cavity walls. They can be made of various materials including wood and PVC.

Door frames are usually standard sizes, but windows are available in numerous shapes and sizes. A sample of door and window frames is shown in Figure 7.19.

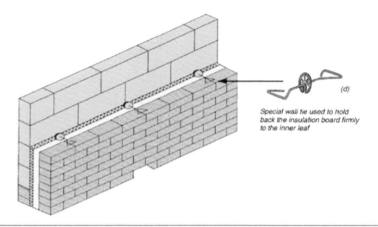

FIGURE 7.17
Partially filled cavity

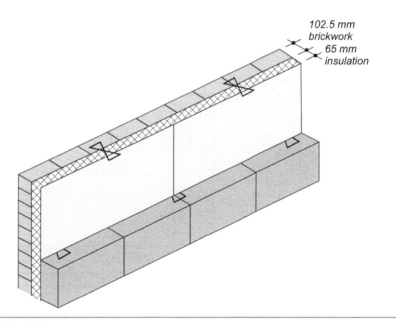

FIGURE 7.18
Totally filled cavity

Construction of cavity walls

Cavity walls commence below ground level unless the work is on a raft construction. Work below ground level has been covered in Level 1, Chapter 10. Please refer back to this chapter if required.

Level 2 deals with cavity walls above ground.

As with all brick and block walls it is essential to set the work out correctly before starting to lay the materials.

The facing bricks should be taken from several packs to avoid banding, but unlike one-brick walls the materials are required for both sides of the wall (Figure 7.20).

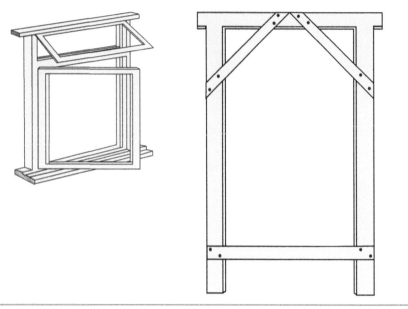

FIGURE 7.19
Door and window frames

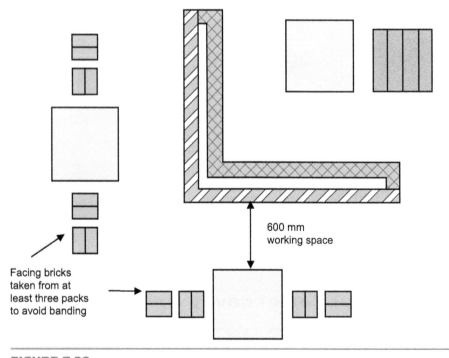

600 mm
working space

Facing bricks
taken from at
least three packs
to avoid banding

FIGURE 7.20
Suggested layout for a cavity quoin

Remember the setting-out details from Chapter 6. Good practice means lifting up the mortar board on blocks to make it more convenient to reach the mortar. Place the mortar boards 600 mm from the face of the wall to allow for adequate working space.

Setting out bonds in cavity walls

When the brickwork of a building approaches ground level it is normal practice to change from common bricks to facing bricks. This changeover normally takes place two courses below finished ground level to avoid commons being shown when the building is complete.

When the facings reach ground level preparation must be made in the bond arrangement for inserting door and window openings at a future height. It is at this stage that the proposed elevation of the building must be carefully studied.

All buildings will have doors and windows built into them and this will lead to a long wall being split up into smaller walls.

It is essential that the bonding is set out at ground level to avoid the problems of broken or reverse bond when the windows are bedded into position later.

Figure 7.21 shows a typical elevation of a building with openings.

Dry bonding

The normal practice is to dry bond the bricks from each end of the wall, placing the 'broken bond' under the door or window opening, where it will be lost for a while.

Setting out facework is usually the responsibility of the supervisor who will, after consulting the architect, determine the detailed bond pattern and the location, if any, of broken or reverse bonds.

Setting out the brickwork is completely different from setting out the building which is done before excavation begins.

One of the main purposes of setting out facework is to create a matching and balanced appearance of bricks, particularly at the reveals on either side of the door and window openings and the ends of the wall.

It is essential that no straight joints occur.

Reveals

The position of all window openings and 'reveal' bricks should be identified when setting out the first two courses (Figure 7.22).

This ensures that perpends can continue unbroken for the full height of the wall.

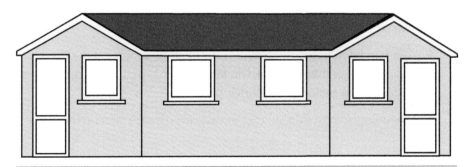

FIGURE 7.21
Semi-detached bungalow

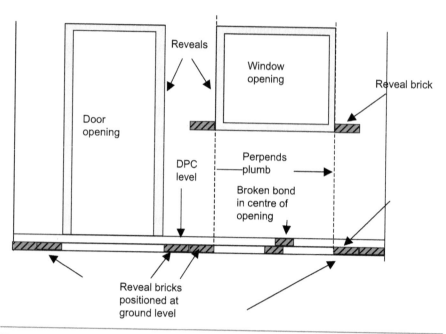

FIGURE 7.22
Bond set out on first two courses

Broken bond

Reveal bricks provide fixed points between which the bonding is set out. The usually short lengths of brickwork between windows offer little scope for adjusting the widths of cross-joints in order to avoid broken bond.

Broken bond can sometimes be avoided by 'tightening' or 'opening' the joints. In doing so, bricklayers should work to the brick size plus a 10 mm joint = 225 mm.

It is far better if broken bonds can be avoided altogether, but where openings and piers do not allow for brick sizes, the broken bonds should be carefully set out. Once their position has been decided, this should be maintained throughout the height of the wall.

As shown in Figure 7.22, the broken bond under the window will disappear once the window is built in.

On no account should a closer be built in the middle of the wall; it should only be placed next to the quoin header.

Broken bond and possibly wasteful cutting can be avoided if the overall length of walls and the widths of doors, windows openings and brickwork between the openings are all multiples of a brick stretcher.

The bonding either side of reveals will also match symmetrically at each course.

In practice, the ideal situation seldom occurs and a satisfactory solution is dependent on bricklaying skills.

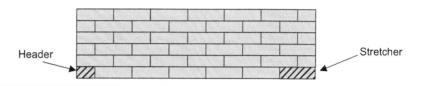

FIGURE 7.23
Reverse bond

Reverse bond

Occasionally it is possible, when setting out facework, to avoid broken bond at the centre of a wall or window pier by 'reversing' the bond.

This means breaking Rule 3 (see Figure Table 6.2 in Chapter 6). The end bricks do not correspond (Figure 7.23).

Erecting quoins in cavity walls

Cavity walls consist of two leaves which in modern construction are usually facework on the outer leaf and blockwork on the inner leaf (Figure 7.24).

Once the corners have been erected and checked for accuracy the remainder of the wall can be run in to a line.

Thermal insulation

When cavity walling was fi used, it was common practice to install air bricks at the top and bottom, at intervals around the whole perimeter of the building, to ventilate this 50–75 mm wide space to remove damp air. Since the 1950s, however, cavity walls have become sealed, with insulating material built in as work proceeds, to improve the thermal insulation value of the cavity space.

Two ways that a bricklayer may be told to install thermal insulation in cavity walling are shown in Figures 7.25 and 7.26.

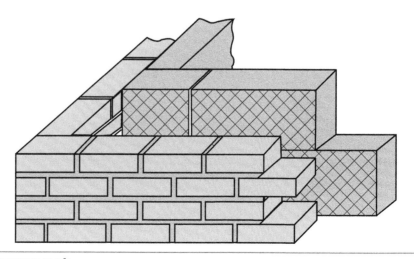

FIGURE 7.24
Cavity quoin

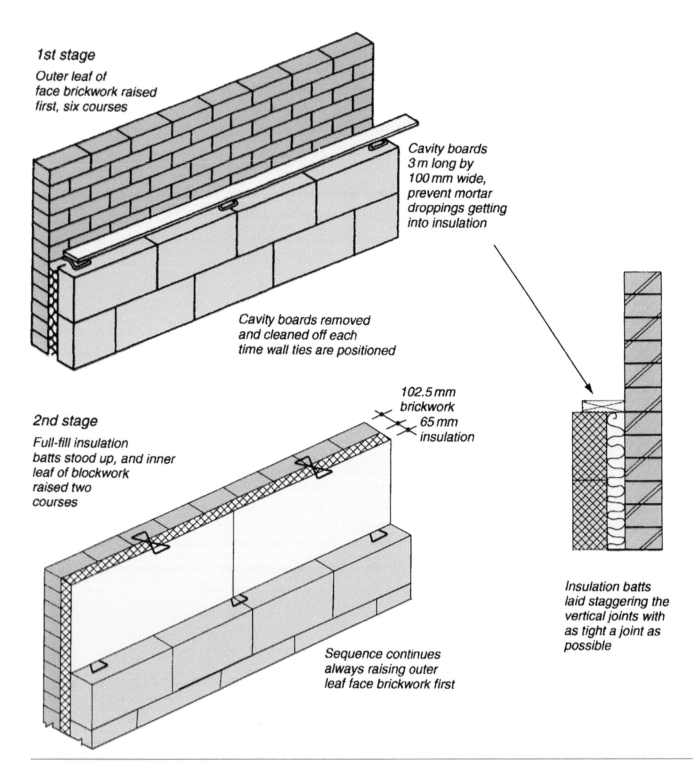

1st stage

Outer leaf of
face brickwork raised
first, six courses

Cavity boards
3 m long by
100 mm wide,
prevent mortar
droppings getting
into insulation

Cavity boards removed
and cleaned off each
time wall ties are positioned

2nd stage

Full-fill insulation
batts stood up, and inner
leaf of blockwork
raised two
courses

102.5 mm
brickwork
65 mm
insulation

Sequence continues
always raising outer
leaf face brickwork first

Insulation batts
laid staggering the
vertical joints with
as tight a joint as
possible

FIGURE 7.25
Fully filled cavity insulation system

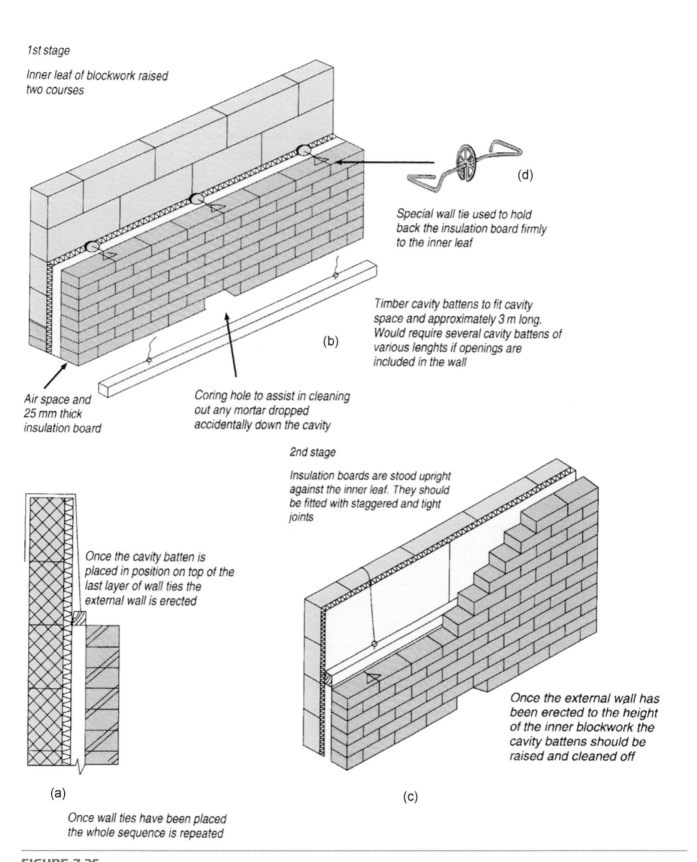

1st stage

Inner leaf of blockwork raised two courses

(d)

Special wall tie used to hold back the insulation board firmly to the inner leaf

Timber cavity battens to fit cavity space and approximately 3 m long. Would require several cavity battens of various lenghts if openings are included in the wall

(b)

Air space and 25 mm thick insulation board

Coring hole to assist in cleaning out any mortar dropped accidentally down the cavity

2nd stage

Insulation boards are stood upright against the inner leaf. They should be fitted with staggered and tight joints

Once the cavity batten is placed in position on top of the last layer of wall ties the external wall is erected

Once the external wall has been erected to the height of the inner blockwork the cavity battens should be raised and cleaned off

(a)

Once wall ties have been placed the whole sequence is repeated

(c)

FIGURE 7.26
Partially filled cavity insulation system

In a fully filled system, flexible fibre batts of insulation completely occupy the cavity space. In a partially filled system, stiffer boards of insulating material half-fill the cavity, but a 25–35 mm air space is retained as well. Both batts and boards are supplied in purpose-made sizes to fit neatly between layers of wall ties and are approximately 900–1200 mm in length.

All insulation batts should be fixed staggering the vertical joints, and butting the vertical and horizontal joints as tightly as possible. Insulation batts can be cut with a sharp knife, but always ensure that the cut is square and forms a perfectly tight joint.

Insulation batts for full-fill insulation are available in various thicknesses and can be made from mineral fibre, which is soft and flexible. Insulation for partial cavity fill is made from more rigid insulation boards such as expanded polystyrene bead board. The boards in partial cavity fill should only be fixed with special wall ties with plastic clips to hold the insulation back against the internal blockwork.

For cavity walling to be effective, wall ties, insulation and cavity gutters must be kept free of mortar droppings as work proceeds. If, owing to carelessness and poor supervision, cavities are not kept clean, then dampness will be able to cross the cavity through porous mortar droppings.

Cavity battens or boards, approximately 3 m long, raised and cleaned off every six courses as work proceeds, are the best way of preventing mortar droppings falling into cavity walling (Figure 7.25). Alternatively, where fully filled cavity insulation is specified, plain battens without lifting wires are used (Figure 7.26).

Coring holes

It is usual to leave temporary openings, called coring holes, in the outer leaf, over all cavity trays. These holes, of one-brick size, are for the removal of any mortar droppings, tools, etc., that have gone past the cavity battens (see point A in Figure 7.27). Cavity ties and trays should be inspected and cleaned in this way at the end of every day's work.

Where possible, however, it is better if one-block-sized coring holes are left out of the inner leaf (see point B in Figure 7.27). This is recommended as it pro- vides a bigger temporary opening. It also avoids the risk of the slight difference in mortar colour that would highlight the coring holes in the finished facework. The coring holes are sealed up when the external scaffolding is removed.

Temporary bedding bricks or blocks in sand at point A or B provide support for those above, and make for easy removal to form the coring holes.

Coring holes should only be regarded as a back-up procedure, and not a substitute for cavity battens or boards.

See also Figure 7.26, which shows a coring hole at ground level

Openings

Leaving a door or window opening in a solid brick wall is a relatively simple job, as it merely requires making allowance for the correct space for a frame, plus a lintel or arch across at the right height.

> **Note**
>
> Always try to keep the cavity clear

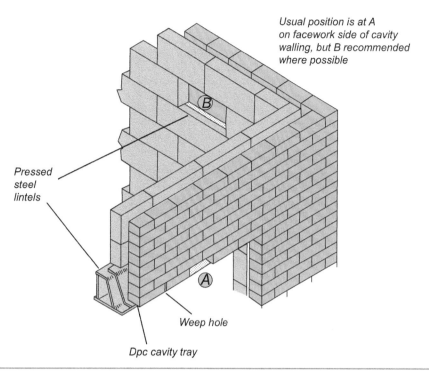

Usual position is at A on facework side of cavity walling, but B recommended where possible

Pressed steel lintels

Weep hole

Dpc cavity tray

FIGURE 7.27
Location of coring holes

Leaving the same opening in a cavity wall, however, is a much more complicated process.

Jambs

The sides of openings are termed jambs, with the face known as the reveal. In cavity walls these could be constructed either square or rebated.

The cavity can be closed at the opening by using a suitable frame or by turning the inner leaf towards the outer leaf.

When the inner leaf abuts the outer leaf a vertical DPC must be inserted to prevent moisture passing through. The vertical DPC should extend into the cavity; 150 mm DPC is advisable when 100 mm blocks are being used.

Detailed provision must be made for dealing with:

- dampness, which may soak through the sides or reveals of any opening (Figures 7.28 and 7.29)

- rain, which may soak through any sill or threshold of an opening (Figures 7.30–7.34)

- rain water, which may run down inside the cavity from above the opening (Figures 7.35–7.37).

Reveals

Reveals require vertical DPC (Figure 7.29).

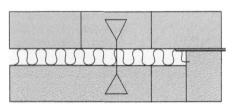

Square reveal with 150 mm
flexible vertical DPC

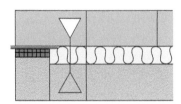

Square reveal with
proprietary cavity closer
with insulated DPC

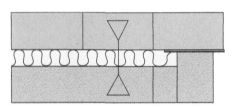

Recessed reveal with 150 mm
flexible vertical DPC

FIGURE 7.28
Plans showing location of vertical DPC at reveals

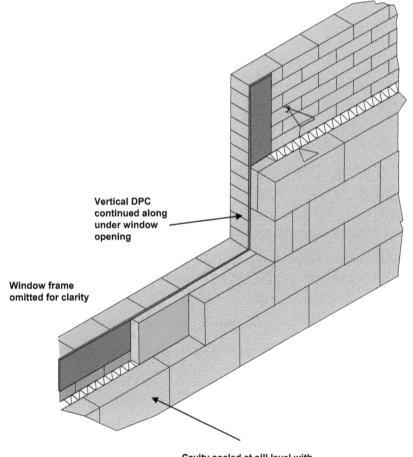

**Vertical DPC
continued along
under window
opening**

**Window frame
omitted for clarity**

**Cavity sealed at sill level with
polystyrene to prevent cold bridging
across the cavity**

FIGURE 7.29
Details showing location of vertical DPC at reveals

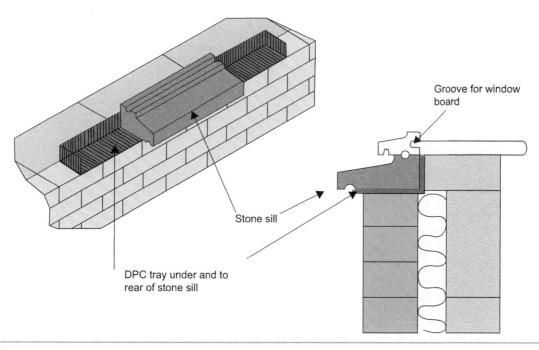

FIGURE 7.30
Construction at timber frame on stone sill showing cavity tray

Sills

The function of the sill is to shed the rain water from the window frame. The sill is very often built into the window frame, but occasionally the frame will not have a sill and the bricklayer will have to construct one.

A DPC tray should be built under sills (Figures 7.30 and 7.31). Sills can also be constructed from creasing tiles (Figure 7.32).

Thresholds

As with window frames, door frames can have purpose-made bottoms fitted during manufacture which are known as thresholds. An example is shown in Figure 7.33.

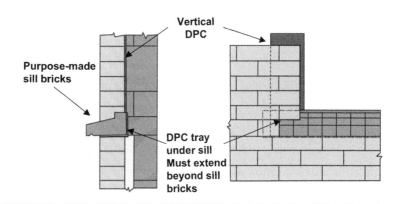

FIGURE 7.31
Purpose-made sill bricks at window opening

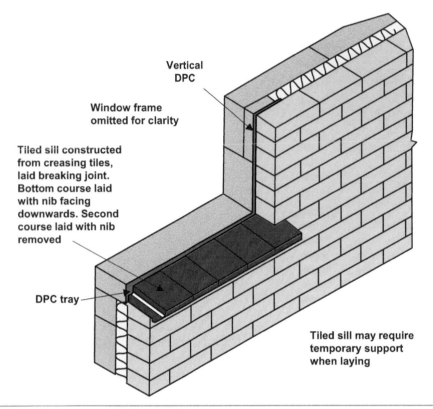

FIGURE 7.32
Construction of tiled sill

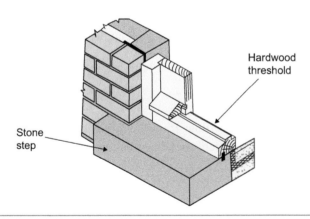

FIGURE 7.33
Door frame with bottom threshold on stone step

Some frames are also available without a bottom and will require the bricklayer to construct a threshold.

Bricks and tiles used for either sills or thresholds should be made from well-burnt bricks able to withstand severe weather conditions.

Engineering quality bricks are most suitable as they are water resistant and have very good wearing properties.

The mortar for this type of work should be equal to the strength of the bricks: one part of cement to three parts of sand.

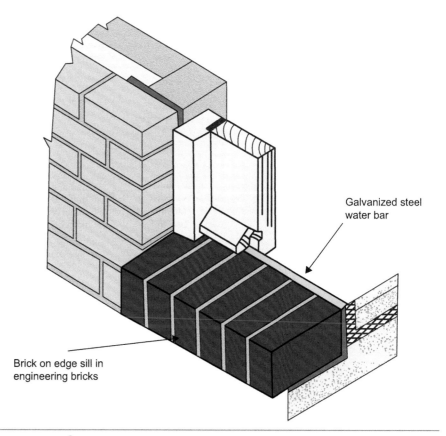

Galvanized steel
water bar

Brick on edge sill in
engineering bricks

FIGURE 7.34
Brick sill details

Bricks on edge are laid sloping to provide a weathered finish. A galvanized steel water bar should be inserted to prevent water blowing under the door. See Figure 7.34 for details.

Damp-proof course cavity trays

Bearing in mind that rain water will penetrate the outer skin of brickwork, trays or gutters of fl bituminous felt must be built into cavity walls to collect the water and prevent dampness from reaching the inner skin of blockwork (see Figure 7.35a, showing a patent steel lintel bridging an opening).

These cavity gutters of DPC material are needed above every opening in cavity wall construction and at every floor level of a multistorey building where cavity walling is used as the external cladding.

Purpose-made stop ends of DPC material (Figure 7.35b) are glued at either end, to prevent run-off into wall insulation and to encourage drainage through weep holes.

Running laps at the end of one roll of DPC and the next, including internal and external angles, must be effectively sealed and jointed, so as to remain watertight. Details are shown in Figure 7.36.

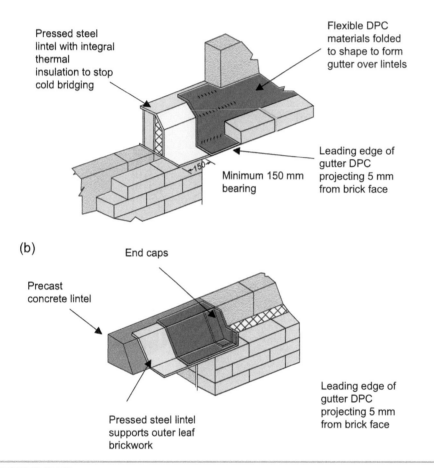

(a)

Pressed steel lintel with integral thermal insulation to stop cold bridging

Flexible DPC materials folded to shape to form gutter over lintels

←150→

Minimum 150 mm bearing

Leading edge of gutter DPC projecting 5 mm from brick face

(b)

End caps

Precast concrete lintel

Pressed steel lintel supports outer leaf brickwork

Leading edge of gutter DPC projecting 5 mm from brick face

FIGURE 7.35
(a) Typical cavity tray or gutter; (b) purpose-made end caps glued to cavity tray

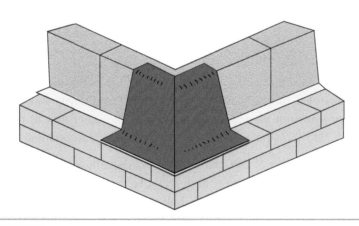

FIGURE 7.36
Purpose-made DPC tray corner unit

Purpose-made polyethylene units may be specified to provide permanent support to running laps in DPC cavity gutters (Figure 7.37).

Rolls of DPC should always be stored on end to avoid squashing and distortion, which would make the material more difficult to flatten out when used. In cold

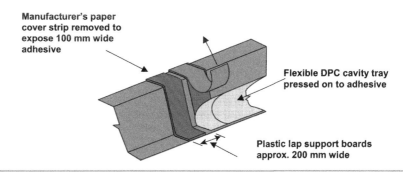

Manufacturer's paper cover strip removed to expose 100 mm wide adhesive

Flexible DPC cavity tray pressed on to adhesive

Plastic lap support boards approx. 200 mm wide

FIGURE 7.37
Patent rigid plastic support board for running laps in DPC trays

weather, the rolls should be stored in a warmer place to make bedding the material easier.

Weep holes

Empty or 'open' cross-joints are left at intervals of 900 mm in the outer leaf of cavity walling, at the level of cavity gutters, to act as permanent drainage points. These weep holes may simply be left as a cross-joint free of mortar, or formed with plastic or nylon fibre inserts.

Both types of inserts are available in a limited number of colours to match mortar joint colour. Where a light colour mortar is specified, open weep holes appear black and will be conspicuous if not in matching locations above and below every window opening. On the external elevations of a building, it is important that the bricklayer takes care to keep the pattern of weep holes under and over openings symmetrical (Figure 7.38).

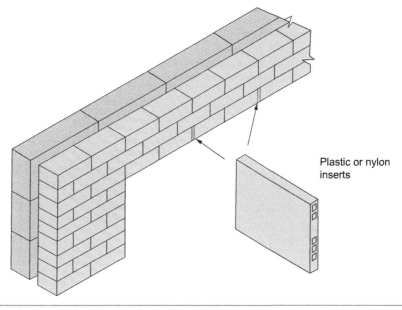

Plastic or nylon inserts

FIGURE 7.38
Position of weep holes and plastic inserts

Cavity details at eaves level

To prevent birds that have entered the roof spaces from nesting in wall insulation, it is usual to close or seal off cavity walls at the top.

Blocks laid flat are a convenient way of doing this. This also spreads the load of a pitched roof between both inner and outer skins of brick masonry (Figure 7.39).

The roof rests on a timber wall plate which has galvanized restraint straps to secure it to the inner leaf.

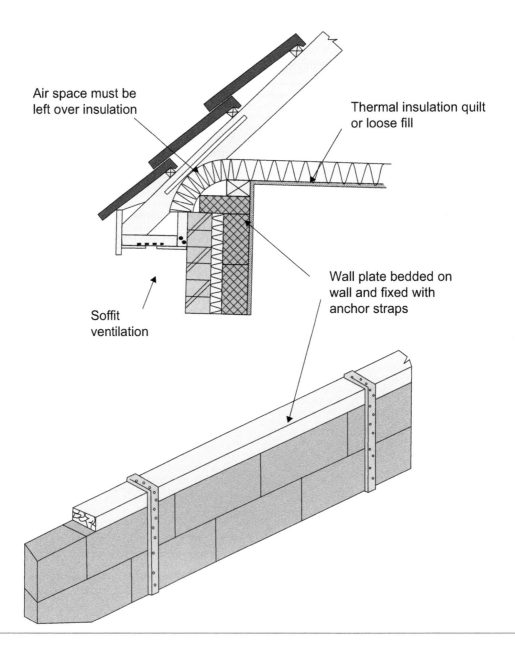

FIGURE 7.39
Sealing cavities at eaves level

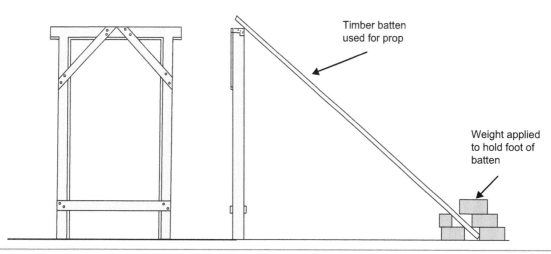

FIGURE 7.40
Door frame support

Openings

Purpose

Openings are formed in cavity walls to allow door and windows to be fitted. Openings could also be produced as access or decorative detail without a frame being fitted.

Frames

Frames may be fitted as the work proceeds or after the wall is completed.

Frames built in as the work proceeds

The bricklayer has to build frames in as the work proceeds and this is often the preferred method for the bricklayer.

The frames can act as a profile, which will reduce the amount of plumbing.

The frame should be placed in position and held in place with timber battens (Figure 7.40). Window frames can be supported in position with a batten (Figure 7.41).

The following factors need to be considered when building in frames:

- Ensure that the horns are cut and splayed back before positioning the frame.

- Before setting the frames in their correct position it is good practice to check the diagonal dimensions to prove the squareness of the frame.

- Check the drawing for the position of the frame and ensure the rebates are facing the right way.

- Set up frames on an even bed, and plumb and level and brace securely.

See Figure 7.42 for details.

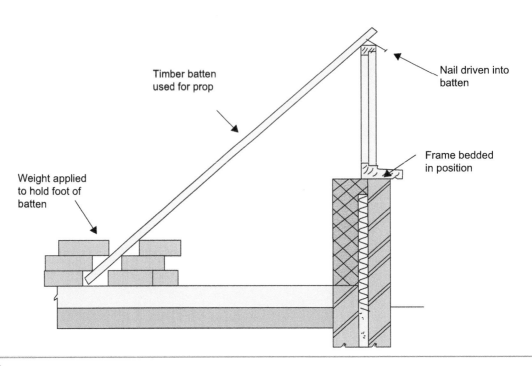

FIGURE 7.41
Window frame support

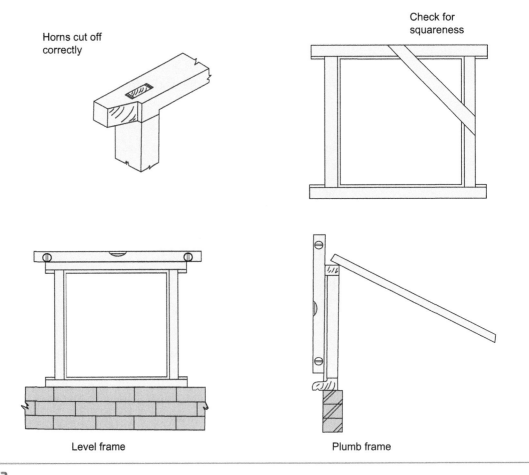

FIGURE 7.42
Setting frames in position

On small frames one batten would be sufficient, but on larger frames two or more may be required.

Fixing frames

The frames should be fastened in position as the wall is being built. This can be achieved using frame cramps (Figure 7.43). If this method is used care should be taken to ensure that the frames are plumb and level, or the joiner would have problems hanging the doors correctly.

Fixing frames after the wall is completed

With the increase in hardwood and plastic window and door frames it has become common practice to leave openings in cavity walls for the frames to be fitted at a later date.

If mistakes are made it can be a very expensive exercise to cut away brickwork to fit new frames.

A threaded lug is screwed into the rear of the frame

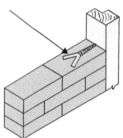

If wood fixing pads are to be used to fasten the frame they should be built in with the grain 90° to the fixing at approximately 4–6 course intervals.

Wooden pad

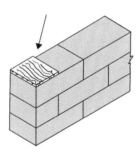

If metal fixing cramps are used they should be built in as the work proceeds

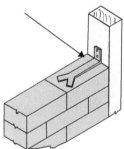

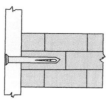

Plastic frame fixers. Timber frame and brickwork are drilled and a plastic plug and screw fixed through the frame into the brickwork

FIGURE 7.43
Methods of fixing frames

Temporary frames

If frames are fitted at a later stage then wooden jigs are built in to provide the correct opening for the frames.

Timber frames are produced exactly to the size of the proposed frame. They should be braced at the corners to make sure that they are square (Figure 7.44).

Pinch Rod

Another method of forming an opening is by using a pinch rod (Figure 7.45). This is a piece of timber cut to the exact size of the opening required and is used to check each course.

As the brickwork is plumbed on each course the pinch rod is held in place to ensure that the opening is not reduced or enlarged.

Protection of frames

Where site traffic has to pass through buildings after the door frame has been fixed it is essential to protect the jambs and where necessary the threshold of the frame (Figure 7.46).

The internal face of the door frame should be protected with timber to prevent knocks which would be expensive to remedy.

If the door has a precast or cast in situ sill then this also needs special protection.

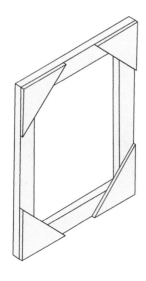

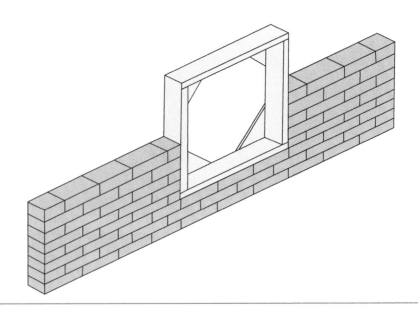

Frames that have to be fixed at a later date will require the bricklayer to form a temporary opening using a temporary wooden jig the exact size of the frame

FIGURE 7.44
Building in temporary frames

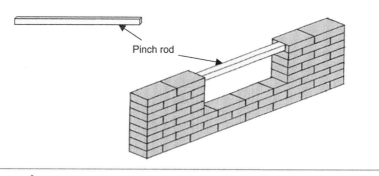

FIGURE 7.45
Use of pinch rod

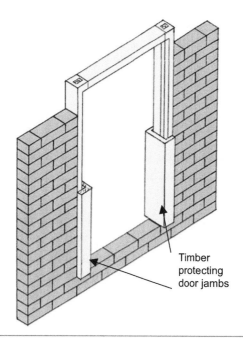

FIGURE 7.46
Protection of frames

Concrete, stone or terrazzo steps can be framed in timber during the remainder of the building work and only removed when cleaning prior to handover.

Where brick steps have been constructed these should also be protected from damage by falling waste materials by covering with timber or similar material.

Junction walls

It is often inconvenient to build junction walls at the same time as the main wall, so there will be a need during construction to make provision for work at a later date.

Indents can be left in the main wall to receive the junction wall later. These indents should be marked out correctly and kept plumb throughout the building of the block wall. The minimum lap at T junctions should be quarter

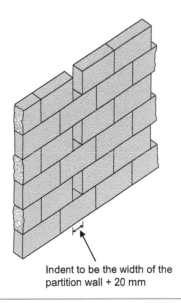

Indent to be the width of the
partition wall + 20 mm

FIGURE 7.47
Indents

lap. An allowance of 20 mm is usually made over the width of the block to allow the blocks on the junction wall easy access into the indent (Figure 7.47).

Another acceptable method of providing a tie to junction walls is metal re-inforcement (Figure 7.48). This is built into every alternate bed joint. This method could help to avoid lots of cutting at a junction of two block walls. This method is recommended for bonding block walls together when both walls are constructed of different block types, especially when the blocks have different shrinkage properties.

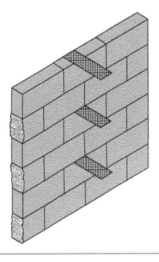

FIGURE 7.48
Reinforcing mesh

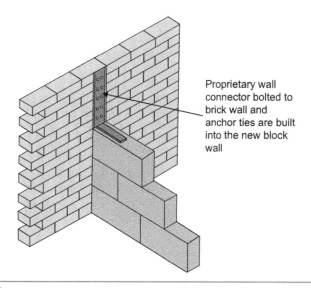

FIGURE 7.49
Proprietary wall connector

Bonding to brick walls

There will be a need to bond block walls to brick walls. This can be done by leaving indents in the brickwork or by fixing a proprietary wall connector (Figure 7.49).

Extensions to cavity walling

It is occasionally necessary to leave part of a building down and, in preparation for its future extension, toothings are formed as the work proceeds. Toothings are shown in Figure 7.50.

The recommended method of erecting a cavity quoin is shown in Figure 7.51.

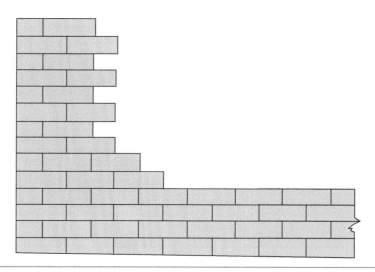

FIGURE 7.50
Toothing out at stopped end

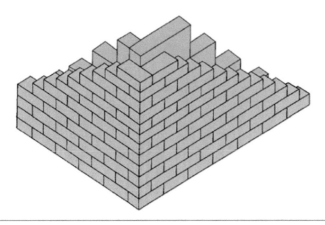

FIGURE 7.51
Cavity quoin

Vertical movement joints

The bricklayer must make allowances, when constructing walls, for the expansion and contraction of the brickwork due to changes in air temperature and/or moisture content of the bricks.

A completely straight joint should be formed throughout the full height and thickness of the wall.

These vertical movement joints should be 10–15 mm wide on the face, and are needed at approximately 9 m intervals to prevent stress from developing in the wall due to expansion. A soft filler of expanded plastic foam strip is built in, to stop mortar bridging the 10–15 mm space.

To reinforce the obvious weakness caused by these unbonded, vertical joints, special stainless steel slip-ties are built in every fourth course as the work proceeds (Figure 7.52) to keep the ends of the walling permanently in line. Standard cross-cavity wall ties must not be used at these points.

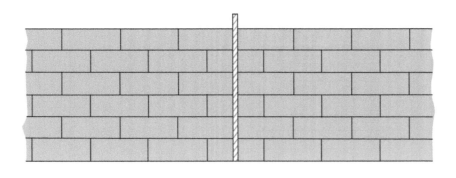

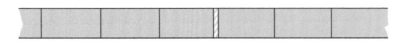

FIGURE 7.52
Formation of vertical movement joint

The expanded plastic foam filler is kept back (or cut back), 12 mm from both faces of the brickwork, so that a sealant-mastic can be applied, to seal these movement joints against rain penetration.

It is very important that vertical movement joints are taken right up through the wall. Vertical movement joints should commence at ground level, because the temperature below ground is practically constant.

Brick corbels

A corbel is the term given to a brick or series of bricks which project from a wall. Corbels are used as a decorative feature or to thicken a wall.

The building of oversailing courses forming corbels – in 56 mm oversailers for the laying and in 28 mm for bond arrangement – is not an easy operation to perform.

In the case of 56 mm corbels, headers should be used wherever possible; if the use of stretchers cannot be avoided, they should be the last bricks laid on that course and should be well bedded, a cross-joint being placed throughout the length of the brick.

In the case of 28 mm corbels, the greatest possible lap must be maintained; if special attention is given to this point, and common sense applied, no difficulty should be experienced in carrying out this work.

Both internal and external straight joints are sometimes unavoidable, but they should be reduced to a minimum and the specified bond should be adhered to, if possible. The final shape of the brickwork must be considered; this will continue to a greater height than the corbel and must be practical and not involve unnecessary cutting.

When a corbel is being built it must be kept well tailed down to prevent overturning; that is, the back bricks must be laid first to bond the previous corbel before attempting to lay the next oversailing course.

The maximum distance permitted by the current Building Regulations is equal to the thickness of the wall immediately below the corbel. See Figure 7.53 for details.

Cutting to rake

A raking cut is required on walls that have to be finished with a sloping end, generally on gable ends.

There are three important points to remember when working on gable ends:

* the kind of finish that is required at the eaves
* how to maintain plumb and level
* cutting the bricks to the required rake.

A typical gable end is shown in Figure 7.54. This shows corbels at the start of the gable.

Note

Compressed fibreboard is unsuitable as a vertical movement joint filler in brick walling. It is too hard, and does not compress easily when the walling expands.

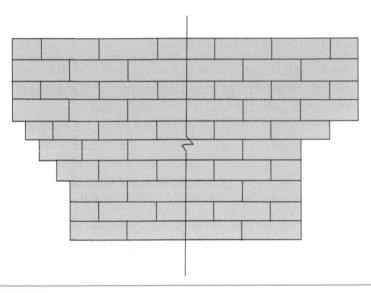

FIGURE 7.53
Two methods of corbelling out brickwork

Gable ends usually start with projected corbels (Figure 7.55). These can be built in brick, tiles or precast concrete.

Corbels require setting out to ensure correct bond.

The joiner will have erected part of the roof on the wall plate and timber laths should be nailed to the rafters at the top and bottom to allow a line to be attached. A gallows bracket could be used at the top, nailed to the ridge board (Figure 7.56).

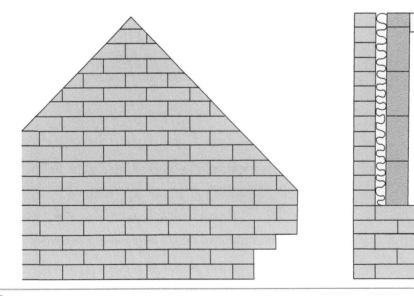

FIGURE 7.54
Typical gable end construction

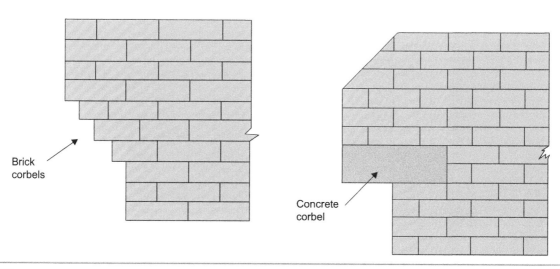

FIGURE 7.55
Types of corbel

The correct position of the timber laths must be ascertained, both for angle of line and for plumb above the main brickwork.

If this procedure is carried out correctly, the remaining brickwork need not be plumbed.

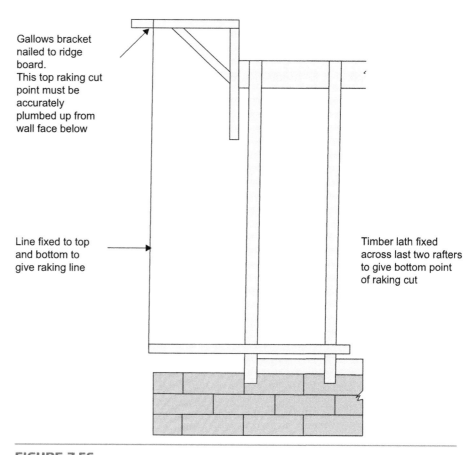

FIGURE 7.56
Setting out for raking cut

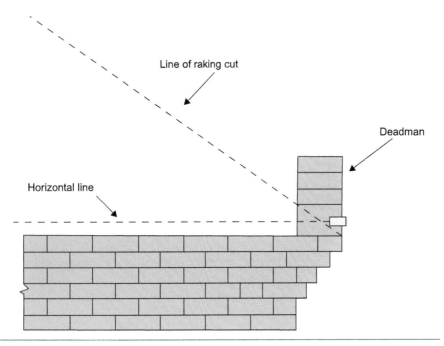

FIGURE 7.57
Use of dead man to support lines

Deadman

Deadman is the term given to a temporary brick pier bedded to gauge and plumbed to the face of the wall. This acts as a corner and allows a horizontal line to be extended beyond the line of the raking cut (Figure 7.57).

A profile could also be used (Figure 7.58).

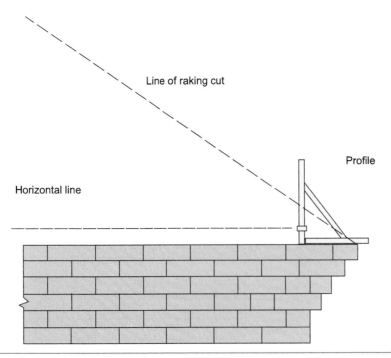

FIGURE 7.58
Use of profile to support lines

The brickwork must be raked back to obtain the correct cut and a fixing for the line and pins.

Procedure for raking cutting

1   Rake back the brickwork below the sloping line and pins (Figure 7.59).

2   Set the sliding bevel at the necessary angle of the raking cutting.

3   Measure the top edge of the cut brick required.

4   Deduct 10 mm from one cross-joint.

5   Mark in pencil the remaining measurement on the brick to be cut, back and front.

6   Draw in pencil the angle of sloping (raking) cut, using the sliding bevel, also back and front, taking care to match the slope on the face.

7   Cut the brick with hammer and bolster.

8   Trim the cut face with a scutch or comb hammer, slightly undercutting if possible to avoid any projections that would prevent the coping or capping from being bedded properly.

9   Carefully bed the cut brick to line and level.

Alternative method, if sliding bevel is not available

1   Rake back and tooth out as before.

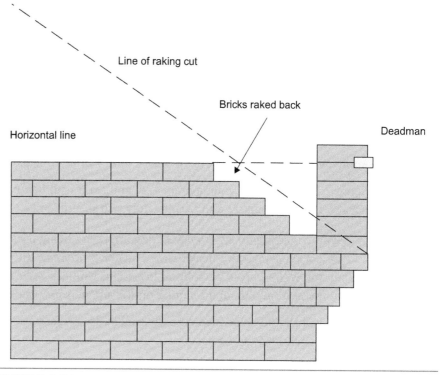

FIGURE 7.59
Brickwork raked back before cutting to rake

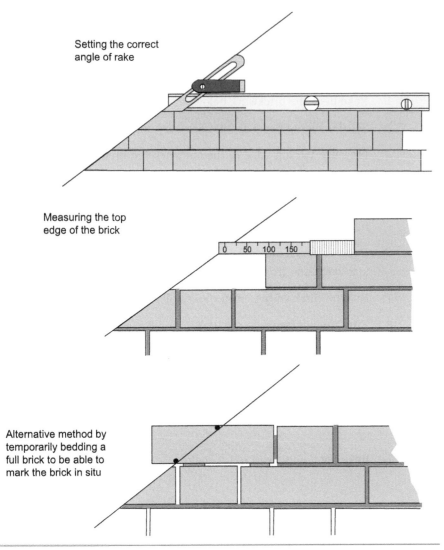

Setting the correct angle of rake

Measuring the top edge of the brick

Alternative method by temporarily bedding a full brick to be able to mark the brick in situ

FIGURE 7.60
Marking the cut bricks

2   Temporarily bed a brick, carefully, to line and level, as shown in Figure 7.60 (bottom drawing).

3   Apply two pencil marks on the brick face to indicate the angle of cut.

4   Remove the brick from its temporary bed, and cut it with hammer and bolster; trim with scutch; and permanently bed the cut brick.

If perforated wire-cut facing bricks or engineering bricks are being used, these will have to be cut on a masonry bench saw. Mark five or six at a time, indicating clearly the waste part of each brick to be cut. Number each one for easy identification, and send them for cutting on the site masonry bench saw.

Each brick to be cut should be bedded onto the wall in its correct position and then marked in pencil ready to be cut.

Use softing to prevent damage and cut with hammer and bolster or mechanical saw.

Procedure for marking and cutting bricks to rake

1 Set the sliding bevel at the necessary angle of the raking cutting.

2 Measure the top edge of the cut brick required.

3 Deduct 10 mm for one cross-joint.

4 Mark in pencil the remaining measurement on the brick to be cut, back and front.

5 Draw in pencil the angle of the sloping (raking) cut, using the sliding bevel, also back and front, taking care to match the slope on the face (Figure 7.60).

6 Cut the brick with hammer and bolster or mechanical masonry saw if available.

7 Trim the cut face with a scutch or comb hammer if necessary to avoid any projections.

8 Carefully bed the cut brick to line and level. See examples of marking in Figure 7.61.

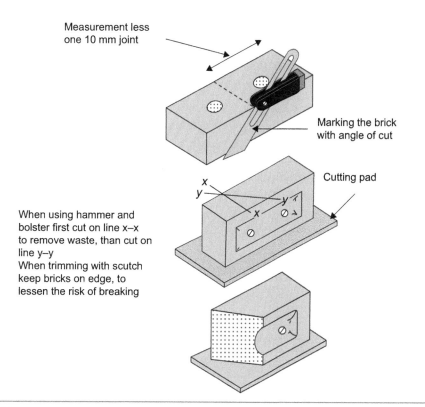

Measurement less one 10 mm joint

Marking the brick with angle of cut

Cutting pad

When using hammer and bolster first cut on line x–x to remove waste, than cut on line y–y
When trimming with scutch keep bricks on edge, to lessen the risk of breaking

FIGURE 7.61
Marking and cutting the bricks

Multiple-choice questions

Self-assessment

This section of the book is designed to allow you to check your level of knowledge. The section consists of revision questions for this chapter. The questions are all multiple choice and have four possible answers. The answers are to be found at the end of the book.

The main type of multiple-choice question will be the four-option multiple-choice question. This will consist of a question or statement, known as the stem, followed by a choice of four different answers, called the responses. Only one of these responses is the correct answer; the others are incorrect and are known as distracters.

You should attempt to answer the questions by choosing either (a), (b), (c) or (d).

Example

The person employed by the local authority to ensure that the Building Regulations are observed is called the:

    (a) clerk of works

    (b) building control officer

    (c) council inspector

    (d) safety officer

The correct answer is the building control officer, and therefore (b) would be the correct response.

Cavity walls

Question 1 What is the main benefit of cavity wall construction?

    (a) decoration

    (b) insulation

    (c) ventilation

    (d) condensation

Question 2 Which of the following is the most common bond used for the external leaf of a cavity wall?

    (a) English bond

    (b) Flemish bond

    (c) stretcher bond

    (d) Dutch bond

Question 3 Identify the type of wall tie shown:

    (a) double triangle

    (b) butterfly

    (c) fishtail

    (d) stainless steel

Question 4 Which is the correct horizontal distance for wall ties in cavity wall construction?

    (a) 300 mm

    (b) 600 mm

    (c) 900 mm

    (d) 1200 mm

Question 5 Name the item of equipment used to prevent mortar dropping down the cavity:

    (a) cavity strap

    (b) cavity trap

    (c) cavity rod

    (d) cavity batten

Question 6 Identify the item of equipment shown:

    (a) window

    (b) door

    (c) jig

    (d) frame

Question 7 Which of the following is the correct minimum width of a cavity?

    (a) 50 mm

    (b) 55 mm

    (c) 60 mm

    (d) 75 mm

Question 8 When would a sliding bevel be used?

    (a) when building in frames

    (b) when cutting bricks to a rake

    (c) when setting out corbels

    (d) when building a quoin

CHAPTER 8

Cladding to Timber, Steel and Concrete Frames

This chapter is about:

- Interpreting instructions
- Adopting safe and healthy working practices
- Selecting materials, components and equipment
- Preparing and erecting brickwork and blockwork to pre-erected structures

The following Skills and Knowledge requirements will be covered:

- Know building types
- Modern methods of construction
- Rapid built technology
- Cladding systems

The following Certificate learning outcomes will be covered:

- Types of walls and their components
- Types of external finishes
- Methods used to insulate cavity walls
- Materials used in cladding pre-erected structures

The following Diploma learning outcomes will be covered:

- Know how to construct masonry cladding to framed buildings

Building information

Definition of cladded walls

The term cladding is used when materials are used to face a pre-erected building.

The standard form of construction for the external walls of brick buildings is cavity walling, which was dealt with in Chapter 7.

This chapter deals with brick cladding to frames of timber, concrete and steel.

Cladding to timber frames is relatively straightforward and is usually restricted to two- or three-storey buildings.

With multistorey buildings, the total loading of all the floors would be too much for normal 275 mm wide cavity walling to support safely. In these situations, a permanent frame of reinforced concrete or structural steel is erected first, before external cavity walling is built around it (Figure 8.1).

Functions of cladded structures

Cladding has several advantages over traditional brick walling:

- The framework can be erected very quickly and possibly waterproofed before the external cladding is required.

- The external brick cladding can be erected without causing any delay to the inside work.

- The whole building will have to be made accessible with scaffolding, so this can be waterproofed, if necessary, to prevent any delay with the brick cladding.

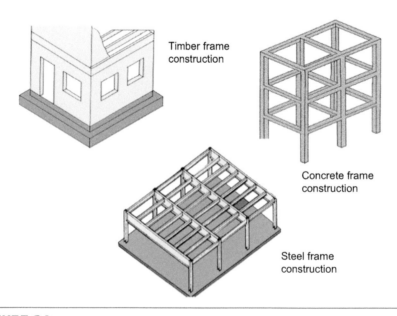

Timber frame construction

Concrete frame construction

Steel frame construction

FIGURE 8.1
Types of frame construction

Design qualities

It is important that the designers of cladded buildings keep the following points in mind:

- Allowances must be made for the thermal movement of the structure and for any movement of the structure due to loading.

- Any movement could cause cracks on the facework and allow dampness in.

Constructional requirements

The brickwork should be securely fixed and supported by the frame. Each floor level will require a horizontal damp-proof course (DPC) tray.

Adopting safe and healthy working practices

Safety legislation has already been covered in Chapter 2. Please refer back to this chapter if necessary.

Particular areas to consider when building cladding to pre-erected frames are:

- personal protective equipment
- manual handling materials
- mechanical tools and equipment
- working at heights.

Resources

This chapter deals with various walling components and materials, including:

- standard bricks and specials
- blocks
- lintels
- damp-proof courses
- door and window frames
- insulation
- expansion and movement jointing material
- wall fixings
- air bricks and ducts
- types of mortar.

Most of these materials have been dealt with in previous chapters. Please refer back to these if necessary.

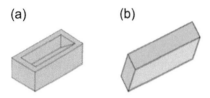

FIGURE 8.2
Types of brick required for cladding: (a) frogged brick; (b) brick slip

Bricks

Two main types of brick are required: normal facings and brick slips (Figure 8.2). Brick slips are required to fill in the wall in front of concrete beams.

Blocks

Some buildings may be designed for brick facings, while others may require faced blockwork. These can also be divided into two categories, load-bearing and light-weight (Figure 8.3).

Lintels

Lintels are required to span across window and door openings.

Concrete or steel lintels can be used inside a building, where they are covered with plaster. Steel lintels are mainly used outside (Figure 8.4).

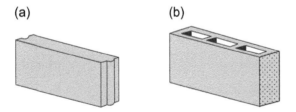

FIGURE 8.3
Types of block: (a) load-bearing; (b) lightweight

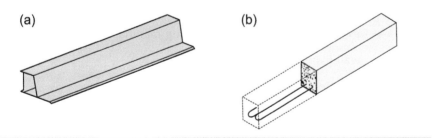

FIGURE 8.4
Types of lintel: (a) steel lintel; (b) concrete precast lintel

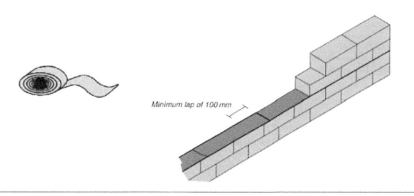

FIGURE 8.5
Damp-proof course

Barriers to damp

The DPC is used as a barrier to rising damp.

The most common type of DPC is the flexible type (Figure 8.5), available in widths of 100, 150, 300 and 450 mm.

The damp-proof materials used vary from flexible sheet material made from bitumen with a hessian or fibre base to rigid material such as slate.

The current Building Regulations specify that the external wall must have a DPC at a height of not less than 150 mm above the finished surface of the adjoining ground.

Cavity trays

The DPC is also used as a cavity tray in cavity walls.

One purpose of providing a cavity between the two leaves of an external wall is to prevent damp penetration.

It is, however, necessary to close the cavity in some situations, particularly around openings for windows and doors (Figure 8.6). In this case, special precautions must be taken to resist damp penetration.

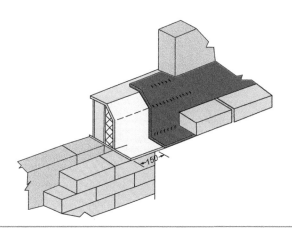

FIGURE 8.6
Cavity tray

It is also necessary to prevent water in the ground from rising up the two leaves of the wall.

Weep holes must be placed at 900 mm centres to allow any water to escape from the cavity.

Components

There are several components that a bricklayer will have to fix as the work proceeds.

Window and door frames

These are built into the walls by the bricklayer as the wall proceeds.

The frames usually have horns on the sill and head. These should be partly cut off, which helps to secure the frame.

Special cramps should also be used to secure the frame every fourth course (Figure 8.7). Door and window frames are also shown in Figure 8.7.

Cavity wall insulation slabs

Slabs are used in cavity walls to provide insulation. They are made of fibreglass or polystyrene, and vary in thickness according to the requirements of the construction (Figure 8.8).

Special wall ties have to be used to secure the insulation slabs.

Expansion joints

When dealing with continuous lengths of cladding it is necessary to allow for expansion of both the bricks and the frame.

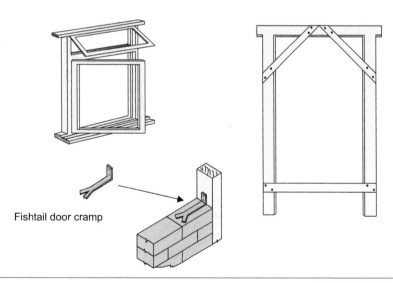

Fishtail door cramp

FIGURE 8.7
Door and window frames

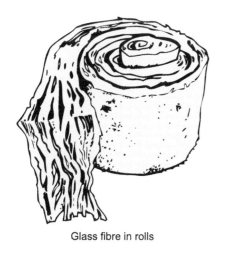

Glass fibre in rolls

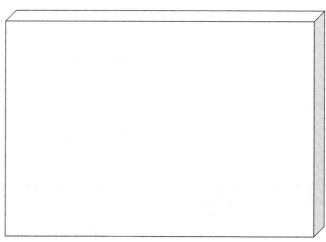

Polystyrene in slabs

FIGURE 8.8
Types of cavity insulation

Open vertical joints should be incorporated every 6–9 m and filled with bituminized foamed polyurethane compound or similar.

Compression joints should also be incorporated; usually at each floor level. These joints should be below the support fixing and are formed by inserting a soft elastic material such as butyl rubber.

Fixings

Brickwork cladding to timber, steel and concrete frame needs to be secured to the frame.

The fixings change according to the material used to construct the frame. A selection of fixings is shown in Figure 8.9.

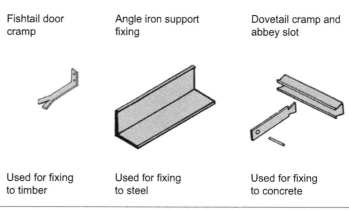

| Fishtail door cramp | Angle iron support fixing | Dovetail cramp and abbey slot |
| Used for fixing to timber | Used for fixing to steel | Used for fixing to concrete |

FIGURE 8.9
Fixings

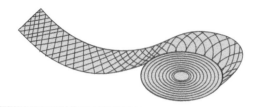

FIGURE 8.10
Flexible mesh

Reinforcement

Several types of reinforcement may be laid into the bed joints of the brickwork to increase the strength (Figure 8.10).

Air bricks and cavity liners

Air bricks are special bricks built into the outer skin of a cavity wall to allow air to pass through to ventilate the cavity. A cavity liner may be inserted, allowing the air to pass through the cavity to ventilate a hollow floor. See Figure 8.11 for details.

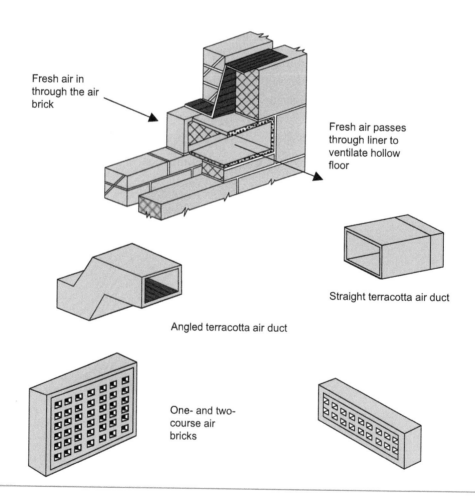

FIGURE 8.11
Air-brick details

The mortar used for cladding should not be too hard or it may cause spalling of the bricks if any movement takes place.

The colour of the mortar should be constant and blend in with the cladding.

Cladding brickwork

The standard form of construction for the external walls to frame buildings is cavity walling.

This means that the bricklayer builds only the external skin of masonry, whether it is brick, block or any other local material. The outer skin is usually 102.5 mm thick face brickwork or facing quality blockwork.

The requirements for cavity wall brickwork and cladding framed buildings are very similar.

Cavity wall brickwork may be load bearing, that is, used to support the load from floors and the roof.

A single skin of 102.5 mm face brickwork is used as brick cladding to framed buildings. The inner skin is formed from timber, steel or concrete storey-height panels, which support floor and roof loads.

All the craft requirements of the bricklayer contained in this chapter apply equally to all applications of cavity walling. As the most common bond is stretcher bond the basic skills gained from Chapter 7 will help with cladding to framed buildings.

The main difference lies in the method of fixing the brickwork to the frame. This will be explained later.

Framed structures

Brick cladding can be used to cover timber, concrete and steel frames.

Timber frames are usually restricted to two or three storeys, while concrete and steel can be used for multistorey buildings.

Timber frames

Brick cladding to low-rise structures is carried out in exactly the same way as for traditional cavity walling. The only difference is the way in which the brick cladding is fi to the timber frame. The foundation, ground works and work up to DPC level are similar to traditional construction.

The timber frame is erected on the concrete slab, protected from rising damp by the horizontal DPC. The whole frame can be erected, including the roof. This can then be waterproofed and work can carry on inside the building while the brick cladding is completed.

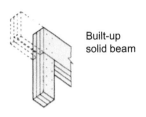

Solid, square or rectangular sections are generally the most economical to adopt

Built-up solid beam

With the larger members it is necessary to form them by combining a number of smaller sections of timber

Plywood beam

This can be done by bolting or nailing several pieces together to form built up sections

Built-up I-section beam

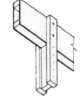

FIGURE 8.12
Timber sections

The prefabricated timber frame is produced off site in a factory under ideal weather conditions. It normally consists of 100 × 50 mm timber with an outer sheathing of plywood or similar material. Insulation is enclosed with an internal lining of plasterboard.

Skeleton frames constructed in timber may be fabricated from solid timber sections, built-up sections or glued and laminated sections (Figure 8.12).

There are three main types of system for assembly on site:

- balloon frame system (Figure 8.13).
- platform frame (Figure 8.14).
- stick system (Figure 8.15).

The balloon system consists of prefabricated frames two storeys high. They arrive on site and are erected off the concrete slab in one stage. The timber joists are suspended from the framework. Gable ends are then erected on top of the two-storey frame if required.

The platform system consists of prefabricated panels of single storey height

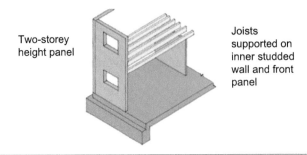

Two-storey height panel

Joists supported on inner studded wall and front panel

FIGURE 8.13
Balloon frame

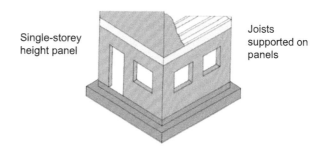

Single-storey height panel

Joists supported on panels

FIGURE 8.14
Platform frame

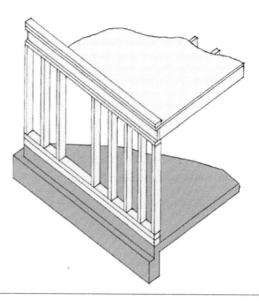

FIGURE 8.15
Stick system: frame constructed on site with individual members

only. These are easier to handle on site. The first floor framework supports the floor joists and the second storey is erected on top.

The stick system is erected on site from individual members and not prefabricated off site. This system takes longer to erect but can overcome small site problems.

Most small-scale timber-framed structures take the form of post and beam construction in which resistance to racking and distortion of the frame under working load is provided by the infill panels.

The roof of a timber frame construction will almost certainly be a roof truss (Figure 8.16). Trussed rafters are fabricated from members of uniform thickness jointed by gluing or nailing, using plywood or, in the case of nailing, punched metal plate gussets.

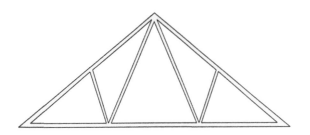

FIGURE 8.16
Roof truss

Brick cladding

Timber wall frames are erected on a foundation of conventional walls or a concrete base or raft to which a sole plate is bolted (Figure 8.17).

The frame is usually assembled as a storey-height frame, in sections around the building. Openings for windows and doors are included in the frame and the frame is covered with a sheathing of timber boards or plywood and a breather barrier.

A sheathing of 9.5 mm plywood sheeting or 25 mm tongue-and-groove boarding is applied to the outside of the frame, which gives it rigidity, provides a face area for fixing back the brick cladding and improves the thermal insulation by sealing spaces between the timber studs.

A vapour barrier is placed on the inner, warm side of the wall frame to exclude

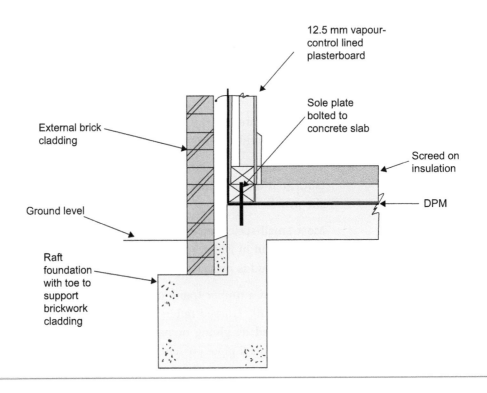

FIGURE 8.17
Detail at ground level

any water vapour passing through the lining which may condense within the cavities. Suitable materials are non-ferrous metal foils, polythene sheeting and especially prepared papers, which are fixed to the framing under the lining; alternatively, aluminium foilbacked plasterboard may be used for the lining. For the barrier to be effective, all joints must be sealed as it must be continuous and imperforate.

A moisture barrier is provided by a building paper stapled to the sheathing or outer faces of the studs with good overlaps at joints. It functions as a second line of defence against penetration of wind-driven rain or snow, or moisture from an external brick veneer.

The brickwork is fastened to the timber frame with galvanized steel cramps fixed every fifth course (Figure 8.18).

Figure 8.19 shows how the cladding is terminated at the eaves and window head, and at sill level.

Figure 8.20 shows a typical section through a timber frame construction including part of the brick cladding to the frame.

Concrete frames

In multistorey buildings, the total loading of all the floors would be too much for normal 275 mm wide cavity walling to support safely. In these situations, a permanent frame of reinforced concrete or structural steel is erected first, then external cavity walling is built around it.

One of the main differences between steel and concrete frames and timber frames is that timber frames provide a total structure, whereas steel and concrete provide only a frame. It is therefore necessary to provide internal walls to both steel and concrete frames. These are known as non-load-bearing brick panel walls.

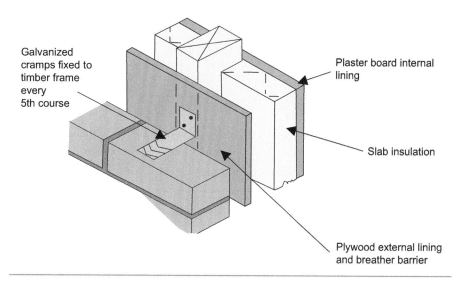

FIGURE 8.18
Fixing brickwork to timber frame

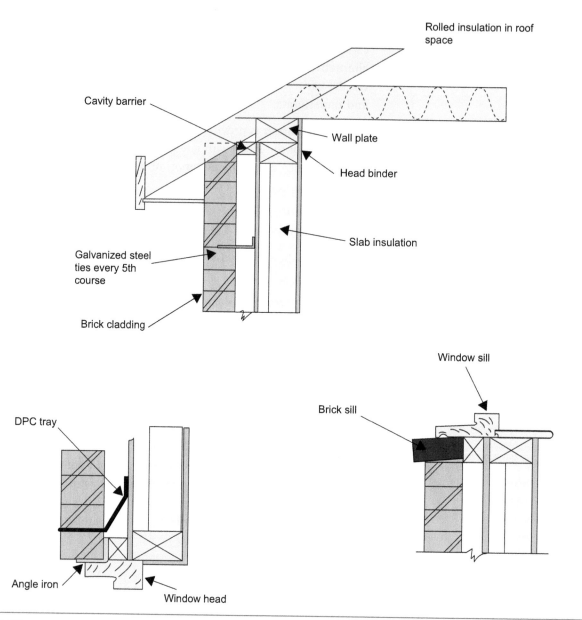

FIGURE 8.19

Details at eaves, window head and sill level

Support for brick cladding

Although the bricklayer builds this cavity walling in exactly the same way as for lowrise structures, when used on multistorey buildings, both leaves of brick masonry must be supported at every floor level. This ensures that permanent pressure does not build up on the brick masonry at ground level. Cavity wall cladding to multistorey buildings is described as non-load-bearing, because it does not carry the weight of floor slabs resting upon it.

Support for the outer skin of brick cladding using reinforced concrete edge beams (Figures 8.21 and 8.22) has generally been replaced by stainless steel angles or brackets (Figures 8.23 and 8.24).

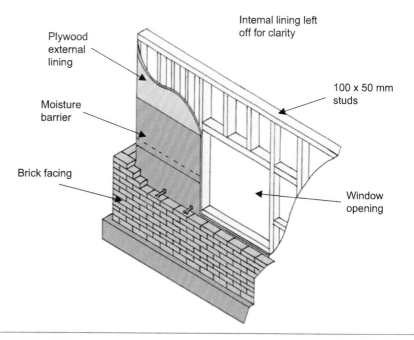

FIGURE 8.20
Details of cladding to timber framework

You will notice that Figures 8.21–8.26 show a soft-joint filler of polyethylene foam strip under the supporting steel angle or concrete toe beams. This very important compression joint must be provided under each supporting angle of stainless steel, or reinforced concrete toe beam, around the external perimeter of a building at every floor level.

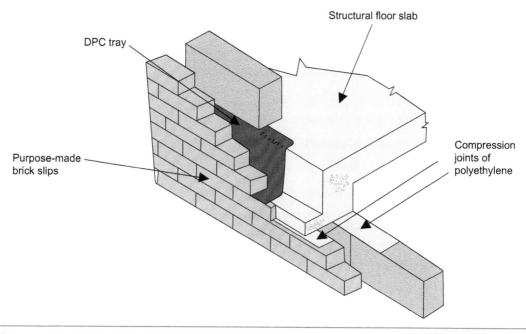

FIGURE 8.21
Support of cavity wall brick cladding on reinforced concrete toe beam

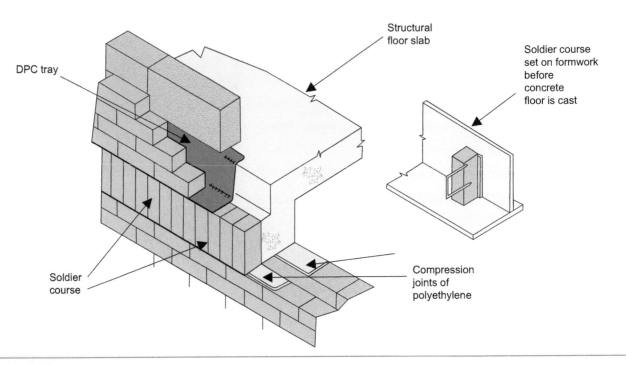

FIGURE 8.22

Outer leaf of cavity wall facework supported on soldier course of bricks, cast in situ with reinforced concrete structural floor slab and edge beam

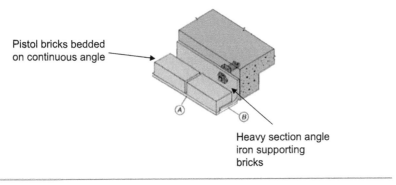

FIGURE 8.23

Heavy section stainless steel angle

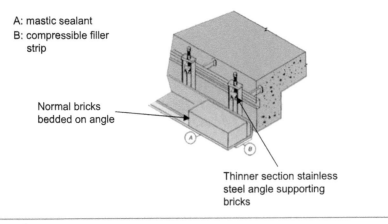

FIGURE 8.24

Thinner section stainless steel angle with adjustable support

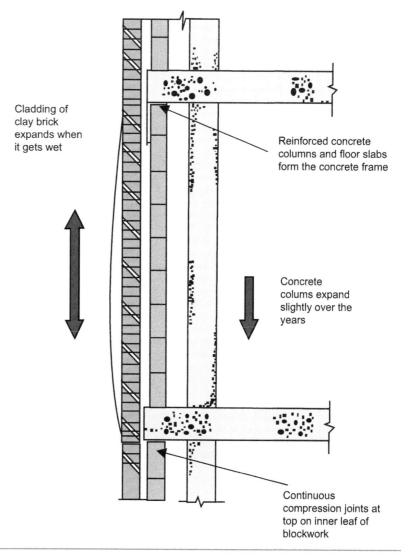

Cladding of clay brick expands when it gets wet

Reinforced concrete columns and floor slabs form the concrete frame

Concrete colums expand slightly over the years

Continuous compression joints at top on inner leaf of blockwork

FIGURE 8.25
The importance of compression joints

The horizontal compression joint, completely clear of mortar, permits different rates of expansion and contraction (up or down) to take place without restriction, between the building and the brick masonry cladding (Figure 8.23).

Unless these horizontal compression joints are provided at every floor level, serious cracking will result when the outer brickwork tries to expand and/or the load-bearing structural frame shrinks slightly over the years (Figure 8.25).

When brick slips are used to cover the concrete frame they need to be fixed to the concrete. Several methods can be adopted; one method, using galvanized steel angles, is shown in Figure 8.26. The galvanized angle is bolted to the concrete frame and the brick slips are bedded into position when the whole building has had time to settle. Fixing the brick slips is one of the last jobs for the bricklayer when working on cladding to concrete frames.

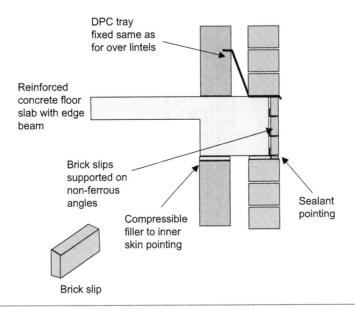

DPC tray
fixed same as
for over lintels

Reinforced
concrete floor
slab with edge
beam

Brick slips
supported on
non-ferrous
angles

Compressible
filler to inner
skin pointing

Sealant
pointing

Brick slip

FIGURE 8.26
Fixing brick slips

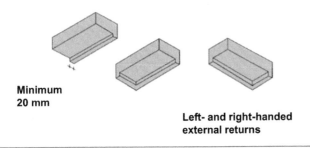

Minimum
20 mm

Left- and right-handed
external returns

FIGURE 8.27
Pistol bricks used to hide thickness of heavy section support angles

Pistol bricks

To disguise the presence of the supporting steel angle within the thickness of a mortar bed joint, the first course of bricks may be rebated (Figure 8.27). These 'pistol bricks' can be produced as bricks of special shape if the manufacturer is given sufficient notice. Alternatively, they can be cut on a masonry bench saw.

Before the importance of horizontal compression joints at every floor level of multistorey buildings was fully realized, great problems developed from the use of brick 'slips', which were used to disguise concrete toe beams.

For these reasons, stainless steel angles or brackets are currently preferred for the support of brick cladding.

Brick panels

Occasionally the architect will design the building to have the concrete beams and columns exposed.

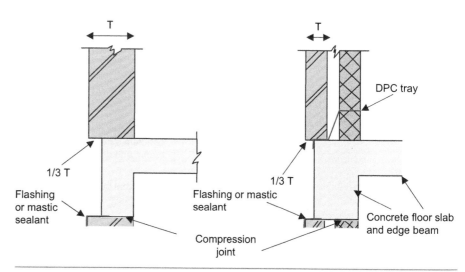

FIGURE 8.28
Maximum permissible overhang

The brickwork is used simply as an infilling panel between the beams and columns. These panels are constructed in the same way as ordinary walls. They need to be adequately supported at floor levels, support their own weight and be weather resistant. Cavity trays and flashings are used to this end.

Brick panel walls are generally constructed with cavities and an insulation block inner leaf to comply with thermal insulation regulations. As in traditional cavity walls, the external leaves are linked together with wall ties.

The wall is supported on the lower concrete floor slab and a DPC incorporated in a similar manner to lintel construction, with the external face of the wall being set forward a small amount beyond the edge of the slab (Figure 8.28). The head of the panel must also be provided with suitable damp-proofing.

Figures 8.21–8.26 showed examples where the concrete frame is covered. Figure 8.28 shows the concrete beam exposed. Where the column is exposed the panel wall is fixed to the side of the concrete column (Figure 8.29).

Special dovetail abbey slots are cast into the columns into which dovetail cramps are inserted to tie the walls to the concrete column. A cheaper method involves casting butterfly wall ties into the concrete column and bending them out when required.

Vertical movement joints
The outer leaf of face brickwork in cavity wall cladding to a framed building will expand and contract owing to changes in temperature and moisture content.

The horizontal compression joints shown in Figure 8.21–8.26 allow up-and-down movements in the brickwork to take place freely.

Allowance must also be made for sideways expansion and contraction, however, so that the outer leaf of brickwork does not cause damage to itself or the structure.

A typical vertical movement joint, with expanded plastic foam joint filler

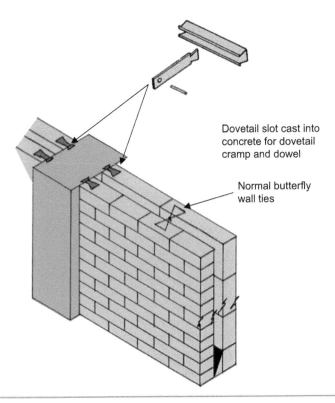

FIGURE 8.29
Fixing to exposed columns

projecting, can be seen in Figure 8.30. This shows 102 mm thick brick cladding to a concrete- framed building.

Vertical movement joints should be provided at intervals between 9 and 12 m for clay brickwork, and not exceeding 7–9 m for sand lime (calcium). Since vertical movement joints form a 'straight joint' weakness in a wall, they must be strengthened with stainless steel slip-ties every fourth course.

Typical details of brick cladding to concrete frames are shown in Figure 8.31.

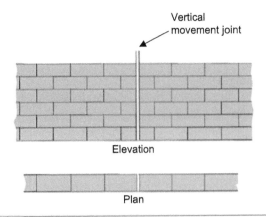

FIGURE 8.30
Brick cladding showing the position of a vertical movement joint

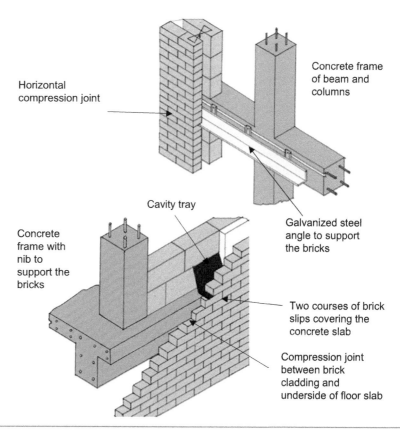

FIGURE 8.31
Various details of brick cladding to a concrete frame

Steel frames

Small building frames are fabricated from sections, either rolled to shape from ingots of hot steel or cold formed to shape from steel strip (Figure 8.32).

Hot-rolled steel sections are standardized in shape and size. 'I'-section girders used as beams are known as universal beams.

Those suitable for use as columns have non-tapered flanges that are much

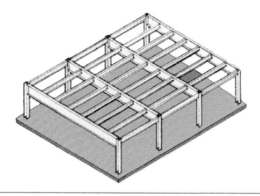

FIGURE 8.32
Typical steel frame

wider relative to the depth of the web than the universal beam, and are known as universal columns.

Hot-rolled members of a steel frame are connected by welded joints or by steel bolts and rivets.

Panel walls to steel frames

Panel walls provide the envelope in skeleton-frame construction.

They are designed to carry only their own weight together with wind loading on their face, and are supported at each floor. Further restraint is provided by the columns and occasionally by the floor above.

Cavity wall construction is used, as for normal cavity walls (Figure 8.33).

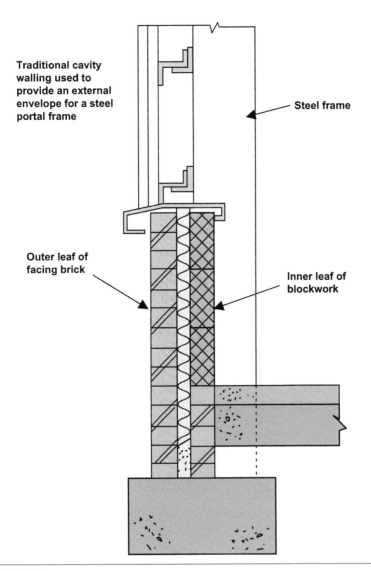

Traditional cavity walling used to provide an external envelope for a steel portal frame

Steel frame

Outer leaf of facing brick

Inner leaf of blockwork

FIGURE 8.33
Brick cladding to a steel frame

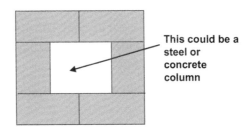

FIGURE 8.34
Two-brick pier in stretcher bond

Brick piers and columns

Occasionally it will be necessary to clad columns and piers, and separate skills for attached and detached piers will be required.

A free-standing brick pier or isolated pier is a pillar of brickwork that stands alone. It can be used to carry the ends of beams over a large span, to support garden gates, etc., or to clad steel or concrete columns.

Piers can vary considerably in size and bond and can be very difficult to build because of the plumbing points.

The bond specified for a particular pier may have to be amended slightly to avoid the excessive cutting required by some bonds. Excessive cutting is expensive in terms of not only materials but also time. When there are many cuts the bricklayer has to take time to ensure that all the cuts are the same, so as to avoid any irregularity in the perpends or the plumbing.

Small section piers are often the most difficult to build because of the limited number of perpends.

If several piers are in line, the piers at either end should be built first and then a line attached to either end pier so that the remainder of piers can all be built in line, both front and back.

Always set the piers out square and to brick sizes.

Hollow piers

Piers are often constructed in stretcher bond and are constructed around the various columns (Figure 8.34). There is no need to fasten the brickwork to the columns as the brickwork will secure itself as it wraps around the column.

Modern methods of construction

There are several new methods of construction being developed to decrease the build time.

The main benefit is that the main structure can be erected and made watertight before the external skin is built.

This allows the inside trades to commence work much sooner.

These latest developments, all classed as rapid construction, are developed to deliver a project within a limited time schedule, approved construction method and meeting the client's satisfaction.

Rapid construction can be achieved through the basic principle of focusing on eliminating waste.

The use of material and workflow in this technology is to provide rapid construction, decrease environmental impacts, increase durability, and reduce on-site labour, resulting in better work zone safety.

Construction firms are transforming how homes are built by building in a different and radically better way that is quicker, cleaner and more predictable than traditional onsite construction techniques.

New developments

Structural insulated panels (SIPs)

Structural Insulated Panels or SIPs for short are sandwich panels consisting of an insulated core that is bonded on each side to orientated strand board (OSB) (Figure 8.35).

SIPs are typically in the external walls and roofs of buildings (Figure 8.36) to create an energy efficient structure in u-values as low as 0.1 W/m2K.

Why SIPS?

- Rapid to install

- Reduced heating bills, excellent U-Values as low as 0.13

- Improved comfort

- Lightweight pre-insulated structure

- Minimal site waste, greener and cost effective

FIGURE 8.35
Typical panel

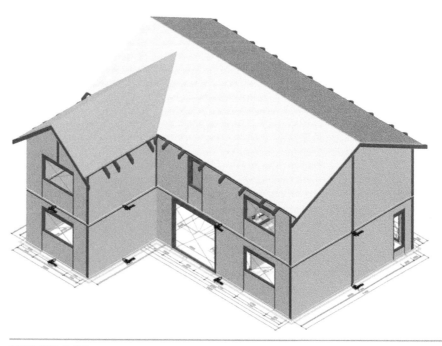

FIGURE 8.36
SIPs detached house

Modular

Modular construction offers three dimensional units delivered with the floor, walls and roof structure assembled offsite (Figure 8.37).

The completion of up to 97% of the build offsite can provide significant benefits across the entire project and remove many site based risks.

Modular buildings are ideally suited to organisations with larger, more complex requirements from their temporary buildings; designed to their exact needs and specifications, however long or short term.

FIGURE 8.37
Modular units

FIGURE 8.38
Prefabricated units

Modular buildings can be adapted to accommodate each client's individual requirements, as well as changing building regulations. Modular buildings can be as technologically equipped as a permanent structure, at a fraction of the cost.

Superior design, delivery and installation practices mean that quality modular buildings can be installed quickly and efficiently, with the minimum of notice or fuss; and modular buildings are designed to last for as long as you need.

Prefabricated buildings

Prefabricated buildings are factory built and designed to provide high quality temporary or permanent accommodation for any industry, within any location in the UK.

Again these are 97% constructed off site and are delivered to the site for installing when the foundation and services are complete (Figure 8.38).

Timber frame

Modern timber frame systems are recognised as the residential construction sectors building material of choice with popularity in timber frame systems now increasing across all construction sectors.

The timber frame system is precision engineered in a controlled factory environment to provide pre-fabricated panels with high levels of insulation, quality and accuracy.

Multiple-choice questions

Self-assessment

This section of the book is designed to allow you to check your level of knowledge. The section consists of revision questions for this chapter. The questions are all multiple choice and have four possible answers. The answers are to be found at the end of the book.

The main type of multiple-choice question will be the four-option multiple-choice question. This will consist of a question or statement, known as the stem, followed by a choice of four different answers, called the responses. Only one of these responses is the correct answer; the others are incorrect and are known as distracters.

You should attempt to answer the questions by choosing either (a), (b), (c) or (d).

Example

The person employed by the local authority to ensure that the Building Regulations are observed is called the:

    (a) clerk of works
    (b) building control officer
    (c) council inspector
    (d) safety officer

The correct answer is the building control officer, and therefore (b) would be the correct response.

Cladding

Question 1 Identify the following fixing used to tie back brick cladding to timber frames:

    (a) tie wire
    (b) fishtail cramp
    (c) fishtail
    (d) butterfly tie

Question 2 What is the maximum overhang for the bricks when the concrete beams are exposed?

    (a) 1/2 thickness of the wall
    (b) 1/3 thickness of the wall
    (c) 1/4 thickness of the wall
    (d) 1/5 thickness of the wall

Question 3 Which of the following is the correct position for a compression joint?

(a) between the top of the wall and the underside of a concrete beam

(b) on top of the concrete beam and under the brick wall

(c) between the top of the brick wall and the bottom of the next brick wall

(d) anywhere in the brick wall

Question 4 Identify the type of fixing for brick to concrete:

(a) fishtail cramp and slot

(b) butterfly cramp and slot

(c) abbey cramp and dovetail slot

(d) abbey slot and dovetail cramp

Question 5 An expansion joint should be produced on long lengths of cladding at intervals not exceeding:

(a) 3 m

(b) 6 m

(c) 9 m

(d) 12 m

Question 6 Identify the type of brick used when cladding concrete frames:

(a) pistol brick

(b) brick header

(c) brick slip

(d) brick stretcher

Question 7 Fixing cramps for cladding should be made of:

(a) non-ferrous metals

(b) steel

(c) galvanized iron

(d) plastic

Question 8 Cladding units are designed to carry:

(a) all loads applied on the structure

(b) no loads applied on the structure

(c) some loads applied on the structure

(d) only live loads applied on the structure

CHAPTER 9
Thin Joint Masonry

This chapter will cover the following City and Guilds units:

- Apprenticeship   Module 3   823
- Certificate   No comparable unit
- Diploma   Unit 203   1.1, 1.2, 1.3, 1.4, 1.5, 1.6, 1.7, 1.8, 1.9
                                              1.10, 1.11, 1.12, 1.13, 1.14, 1.15, 1.16
                                              2.1, 2.2, 2.3, 2.4, 2.5, 2.6, 2.7, 2.8, 2.9
                                              2.10, 2.11, 2.12, 2.13

This chapter is about:
- Interpreting instructions
- Adopting safe and healthy working practices
- Selecting materials, components and equipment
- Preparing and erecting thin joint block masonry systems

The following Skills and Knowledge requirements will be covered:
- Methods used to build thin joint systems

The following Certificate learning outcomes will be covered:
- No comparable unit

The following Diploma learning outcomes will be covered:
- Know how to construct thin joint masonry to framed buildings
- Be able to construct buildings using thin joint masonry

Thin joint masonry information

Continuing research has enabled leading block manufacturers to develop a thin joint mortar which in conjunction with a larger block size enables internal walls to be constructed more quickly.

This system is an alternative to using 10 mm sand and cement mortar bed and cross-joints.

It is essential that all parties have access to various sources of up-to-date information. It is also very important that the student understands where information can be found.

Information is available from various manufacturers, many of whom will provide on-site demonstrations.

Manufacturers' information

Any efficient office should have up-to-date information regarding new and existing products from the various manufacturers who produce materials of interest to them.

Many manufacturers produce technical information which is free for the asking. These papers, brochures, leaflets, etc. (Figure 9.1), should be stored in the office in an easily retrievable system.

Technical information can be produced in several formats. When items of equipment are purchased they will come with a manufacturer's information sheet.

The information may be:

- operating instructions – how to use the item
- safety guidelines – power supply, personal protective equipment (PPE) to be worn and recommended checks
- technical information – mechanical details and possible outputs.

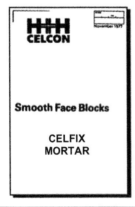

FIGURE 9.1
Manufacturer's literature

Normal hand tools are not usually provided with manufacturer's instructions. Most textbooks explain hand tools in detail and the safety procedures to follow while using them.

Information on setting-out equipment is provided by the manufacturer to ensure that the item of equipment is used correctly.

Adopting safe and healthy working conditions

It is essential to wear the appropriate PPE when laying thin joint masonry, in the same way as for normal masonry.

When the thin joint system is used it is necessary to cut the blocks accurately to maintain the thin joint. The traditional method of bolster and club hammer will not give an accurate cut, so a saw or mechanical method should be used.

When components are cut by machine the equipment must be set up correctly and safely and the operator must have had sufficient training and acquired the required certificate. The appropriate PPE must be worn.

The main type of hand-held equipment for cutting building components is the angle grinder or disc cutter. A table version of a disc cutter is used in workshops and on building sites when there is a great deal of cutting to be carried out.

Please refer back to Figure 6.1 in Chapter 6 for further information.

Resources

As with any new system there will be several new materials to become familiar with. It is essential to practise using these new materials and methods before starting a new project.

Blocks

These are available in various shapes and sizes to cover all types of situation.

Blocks for thin joint masonry have sharp and clear arrises and excellent dimensional tolerances and are produced with a plain face.

Standard blocks are available from 200 mm upwards with a face size of 610 × 140 mm and are designed for solid wall and foundation construction.

Blocks are also produced in various thicknesses. They are suitable for the construction of solid and cavity walls and are available with a face size of 610 × 215 mm.

The blocks are produced to a tolerance to allow them to be used in thin joint systems and all cut blocks need to be cut to the exact size, allowing only 3 mm for the joints.

Remember

It is the employer's responsibility to ensure that you are correctly trained to use power tools. The item of equipment provided must be maintained and safe.

Employees, in turn, have a duty to inform their employer of any work situation that presents a risk to themselves or their workmates.

A masonry saw or mechanical saw may be used. Do not use a club hammer and bolster chisel as this would lead to joint widths greater than 3 mm.

Jumbo blocks are available in various thicknesses with a face size of 610 × 270 mm. They are suitable for the construction of cavity walls. These larger format blocks can increase the speed of construction while improving the thermal performance of the building by reducing cold bridging through the mortar. See Figure 9.2 for details.

Mortar

The key to all thin joint systems is the mortar. The mortar is designed specifically for use with the special blocks.

The dry mix should always be added to the water (approximately 5–6 litres of clean water per bag) and mixed until a thick, smooth consistency is achieved.

Thin joint mortar

A thin joint mortar has been developed and is supplied as a dry premixed powder in 25 kg bags (Figure 9.3). This replaces the traditional 10 mm sand and cement mortar.

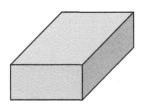

Solid wall blocks
Thicknesses from 200 mm
upwards
Face size = 610 x 140 mm

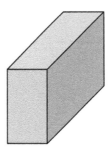

Cavity wall blocks
Various thicknesses
Face size = 610 x 215 mm

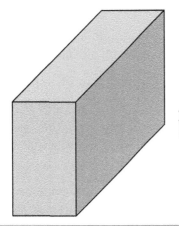

Jumbo blocks – various
thicknesses
Face size – 610 x 270 mm

FIGURE 9.2
Types of block

FIGURE 9.3
Bagged dry mix

A special scoop is used, which creates a joint thickness of 2 mm.

The mortar will remain workable in the bucket or tub for a number of hours, but reaches initial bond strength on the block within 10–20 minutes and full design bond strength within 1–2 hours.

Tools and equipment

Before setting out any work the equipment should be carefully checked for accuracy. New tools and equipment may need to be purchased and practised with. A selection is shown in Figure 9.4. Other items that may be required can be found in Figure 9.5.

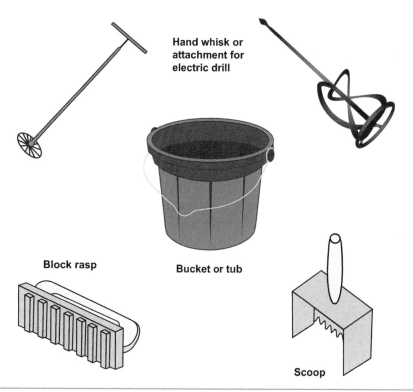

Hand whisk or attachment for electric drill

Block rasp

Bucket or tub

Scoop

FIGURE 9.4
Main tools and equipment required

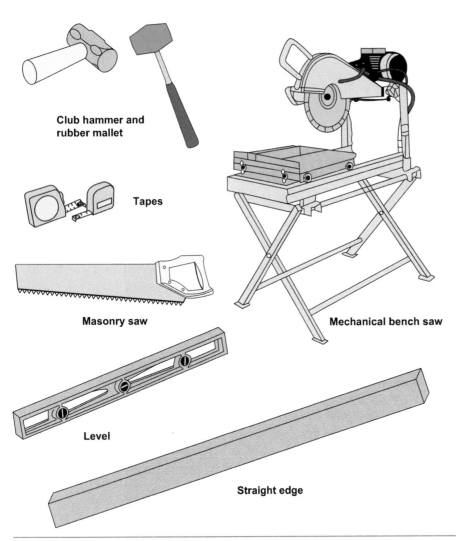

Club hammer and rubber mallet

Tapes

Masonry saw

Mechanical bench saw

Level

Straight edge

FIGURE 9.5
Other tools and equipment required

Thin joint blockwork

A revolutionary system of blockwork, the 'thin joint system', introduces a new material called aircrete. Aircrete blocks combine thermal and sound insulation qualities with load-bearing capabilities.

Aircrete blocks are also:

- highly resistant to the passage of water
- fire resistant, including surface spread
- frost resistant, with no reduction in freeze–thaw conditions
- resistant to sulphate attack up to class 4 soil conditions.

The thin coat system is in use throughout Europe.

Benefits of the thin joint system

Speed:

- reduced construction time

- follow-on trades can start sooner

- speed and ease of fixing secondary insulation

- internal work can proceed while external work is being constructed.

- Quality:

- improved thermal insulation

- airtight construction

- reduction in site waste

- improved stability

- accuracy of walls allows for thin coat plaster finishes

- increased bond strength.

- Flexibility.

Speed

The system enables the structure of a building to be constructed more quickly and to a better quality than traditional construction, allowing follow-on trades to start work sooner in a weatherproof environment, while retaining the flexibility of on-site construction.

The thin joint system allows construction times equivalent to off-site system build solutions.

The system gives increased bond strength together with ease of mixing and laying the mortar with precision-made blocks.

Quality

The thin joint system leads to improved quality of thermal insulation, stability and accuracy, and a reduction in waste.

Flexibility

As with traditional building methods, the construction is carried out on site. The system is easy to extend or adapt as the work proceeds.

Applications of thin joint systems

Thin joint systems can be used in the following constructions:

- cavity walls (inner and outer skins)

- solid external walls

- semi-exposed conditions

- partitions
- separating walls
- foundations
- industrial and commercial buildings.

Cavity walls

Cavity walls in thin joint systems are built very differently from traditional block and brick cavities.

The inner skin can be completed without the insertion of traditional wall ties – these being too thick. This leaves the inner block wall clean and free from dangerous protruding ties.

When the external skin is built the insulation can be inserted and special helical wall ties can be driven into the thin joint system at the appropriate level.

The inner leaf can be built first, taking the outer leaf off the critical path of the building programme.

Solid external walls

The superstructure of a dwelling can often be completed in 2–3 days.

The use of thin joint systems increases the U values by up to 20 per cent compared with conventional joints in solid walls, so the wall can meet thermal regulations without the need for additional insulation.

Partitions

Excellent sound and fire insulation is provided by the thin joint system. With the speed of the build and mortar setting more quickly then traditional mortar, it is possible for the walls to be finished, i.e. plastered, without delay.

Separating walls

Separating and flanking walls constructed with the thin joint system satisfy the sound insulation requirements of the current Building Regulations.

Building thin joint systems

It is assumed that the candidate has previous knowledge of traditional blockwork.

The approach to using the thin joint system is different from that for the sand and cement system. As mentioned previously, the blocks are specially made to a tolerance to allow them to be used in thin joint systems with a 2 mm joint. The stability of the walls is greater than with sand and cement; therefore greater height can be reached with very little or no support required.

First course bedded, on the DPC, in traditional mortar, using 10 mm joints to provide a level course

Sanding board should be used to smooth out any discrepancies

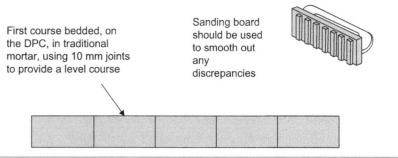

FIGURE 9.6
First course should be perfectly level

Whichever method of foundation has been used under the thin joint system the first course of thin joint blockwork is very important. With traditional foundations the first course is usually laid onto the damp-proof course (DPC) using traditional sand and cement mortar to achieve a level datum to work to.

The first course is the most important, as height cannot be gained or reduced with thin joint mortar as easily as with traditional mortar. This course should be allowed to set overnight. If it is not perfectly level; any irregularities can be sanded down using a sanding board or rasp (Figure 9.6).

If the first course is laid on a level concrete raft, the thin joint system can be used, ensuring that it is true to level and plumb. It is essential that the blocks are laid accurately, true to level (±1 mm) and vertical.

Once the first course has been laid, allowed to set and made perfectly level, the quoins can be erected as for normal blockwork, keeping to half-bond where possible. The recommended method is to lay three courses at a time (Figure 9.7).

Building with the thin joint system is a very fast and easy process. The skills required to build with the thin joint system are very similar to those for traditional blockwork, but regular checks must be made for level, plumb and line. It is difficult to make any adjustments to height as the work proceeds.

Block quoin erected as normal except to a much tighter gauge

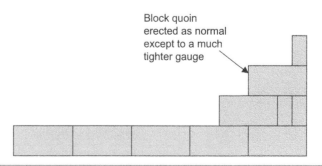

FIGURE 9.7
Erecting a thin joint quoin

Before applying any mortar ensure that all surfaces are clear of loose materials. Avoid using wet blocks as this slows down the setting time.

Plumbing blockwork

Take care when plumbing thin joint blockwork as the wall rises more quickly and the mortar sets more rapidly than when using traditional blockwork.

On no account tap the blocks sideways – this will only result in the bed joint opening up on one side. Always plumb the wall by tapping down on the high side of the block (Figure 9.8).

Laying to line

Again this procedure is the same as for normal blockwork but the gauge is much less: 2 mm joints rather than 10 mm (Figure 9.9).

Bonding

Half-bond should be used as much as possible. When wall lengths do not work exactly to block lengths then broken or reverse bond should be used.

Broken bond

Broken bond may have to be adopted when wall lengths are not equal to block sizes. Most block walls will require some use of broken bond.

It is important never to use cuts less than half a block as this can create a weak plane in the wall and if drying shrinkage occurs it may result in cracks appearing in the wall (Figure 9.10).

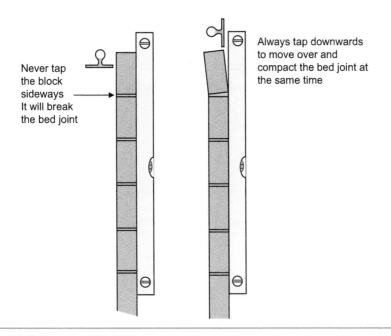

FIGURE 9.8
Plumbing thin joint blockwork

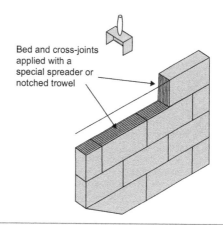

Bed and cross-joints
applied with a
special spreader or
notched trowel

FIGURE 9.9
Laying to line

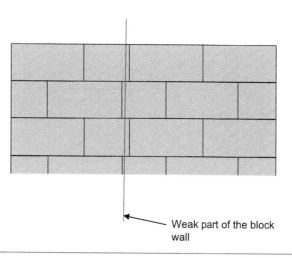

Weak part of the block
wall

FIGURE 9.10
Broken bond

Blocks can be cut with a club hammer and bolster chisel, a masonry saw or a mechanical bench saw (Figure 9.5). Try to make use of any available special blocks before starting to cut. Always try to achieve a neat cut that retains the 10 mm cross-joint. Joints that are too large can again result in cracks appearing in the wall due to drying shrinkage.

Broken bonds of less than half-lap can be avoided using blocks (Figure 9.11).

Reverse bond

In reverse bond the blocks at each end of the wall are different. The front elevation in Figure 9.12 shows reverse bond being used.

When cut blocks are being used it is more economical to use three-quarter, half- and quarter blocks provided by the manufacturer. Cut blocks should always be of the same material and must be accurately cut and smoothed down.

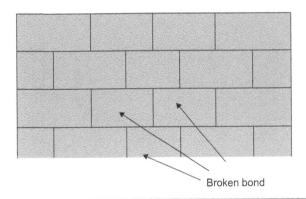

Broken bond

FIGURE 9.11
Broken bond

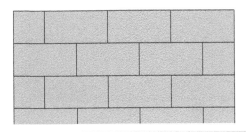

End block
different to
opposite
end block
to avoid
broken
bond

FIGURE 9.12
Reverse bond

Junction walls

It is often inconvenient to build junction walls at the same time as the main wall, so there will be a need during construction to make provision for work at a later date.

Indents can be left in the main wall to receive the junction wall later. These indents should be marked out correctly and kept plumb throughout the building of the block wall. The minimum lap at T junctions should be quarter lap. An allowance of 20 mm is usually made over the width of the block to allow the blocks on the junction wall easy access into the indent (Figure 9.13).

Another acceptable method of providing a tie to junction walls uses flat strip shear ties (Figure 9.13). These are built into the main wall on every alternate bed joint. If the inner leaf has been completed before the outer leaf then helical ties can be driven through the inner leaf into the junction wall by at least 50 mm (Figure 9.14).

These methods can help to avoid a lot of cutting at a junction of two block walls.

Cavity walls

The main benefit of constructing the inner blockwork leaf first is that a weathertight environment is created in which follow-on trades can work.

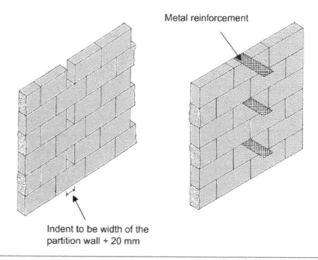

Metal reinforcement

Indent to be width of the
partition wall + 20 mm

FIGURE 9.13
Junction details

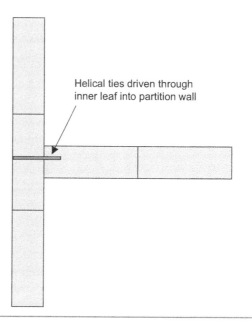

Helical ties driven through
inner leaf into partition wall

FIGURE 9.14
Pinning junction wall

The inner leaf is erected by building corners and running in as for normal blockwork. The mortar bed should remain workable for 6–9 minutes and sets within 10 minutes. Any minor adjustments should be made as soon as possible.

Cross-joints can be applied with the same equipment or the blocks could be dipped into the mortar to apply the cross-joint.

As laying proceeds, blocks should be pressed firmly against the mortared face of the preceding block while being lowered onto the mortared bed below (Figure 9.15).

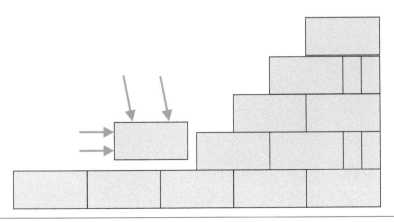

FIGURE 9.15
Laying thin joint blocks

To assist in levelling, a rubber hammer or mallet can be used to tap the block into position, especially to close the perpend joints, and reduce mortar thickness. This will also help to ensure that all the joints are fully filled.

Any excess mortar should be removed from the face of the blocks.

Always stack blocks adjacent to the wall being built for increased productivity. Building to profiles is recommended to improve the accuracy of level, line and speed (Figure 9.16).

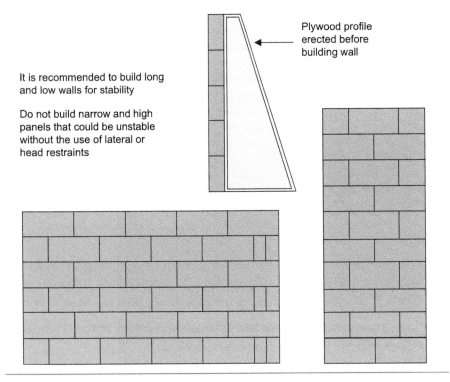

FIGURE 9.16
Maintaining stability

Wall ties

Because of the difference in gauge traditional wall ties cannot be used.

If the inner wall is erected before the outer leaf then wall ties are not required. This is an added safety feature as no wall ties will be projecting during the erection of the outer leaf. The insulation and cavity wall ties can be fitted when the external leaf is erected.

One of the main differences from conventional building methods is the height to which the wall can be built. The inner leaf can be built to wall plate level and the inner work can continue while the outer leaf of the cavity wall is being built.

Wall ties are placed in the normal positions, but owing to the bed joint not coinciding, special helical wall ties (Figure 9.17) are driven into the face of the blocks.

Insulation

Insulation can be placed against the inner blockwork and the wall ties driven through the insulation. If partial insulation is used, special clips are inserted as normal. It is important to ensure the block wall has set and is stable before driving in wall ties.

The benefits of thin wall system can be greatly enhanced if jumbo block units are used. These are available as blocks 440 mm long and 430 mm high, although sizes vary between manufacturers.

The current Building Regulations require that heat loss through mortar joints must be taken into consideration when calculating the U value for walls. The use of thin joint masonry minimizes the effect of heat loss as a result of the 2 mm bed and cross-joints.

If the inner leaf has been completed before the outer leaf the cavity is kept cleaner and also the insulation is cleaner.

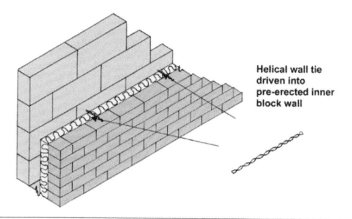

Helical wall tie
driven into
pre-erected inner
block wall

FIGURE 9.17
Helical wall tie

Movement joints

Provision for movement should be considered at the design stage.

Walls in excess of 6 m long should be designed as a series of panels separated by movement joints at not more than 6 m centres. The position of the first movement joint should be within 3 m of a fixed end or corner.

Wall panels are built butting against the next panel, leaving a dry joint or a 10 mm gap that can be filled with fibreboard.

Where lateral stability demands continuity across the joint, proprietary flat strip shear ties should be set in at 450 mm maximum vertical centres (Figure 9.18). These should be set parallel to the plane of the wall, allowing the plain end to become debonded if movement occurs within the blockwork panel, thus creating a movement joint.

Reinforcement

For areas where high stress is expected, such as openings or where blocks are under concentrated imposed loads, horizontal bed-joint reinforcement is recommended. It should be of adequate length to distribute stresses to nearby movement joints or into adjacent blockwork on each side of the opening.

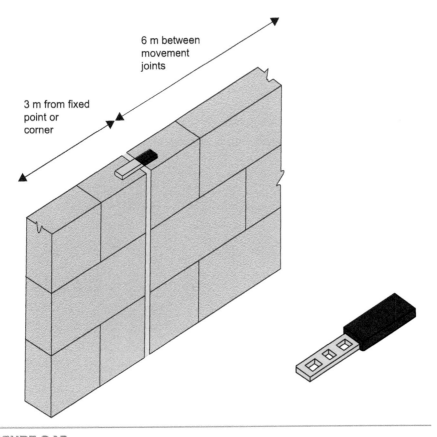

FIGURE 9.18

Movement joints and shear tie

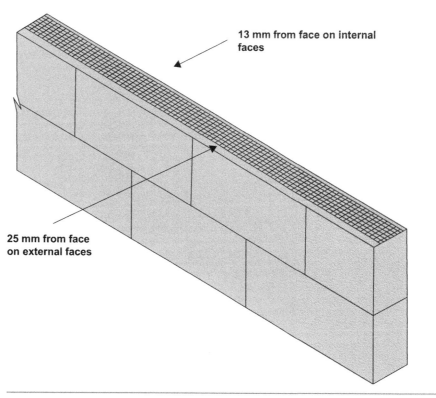

13 mm from face on internal faces

25 mm from face on external faces

FIGURE 9.19
Reinforcement

Reinforcement should normally have a mortar cover of a minimum 13 mm for internal faces and 25 mm for external faces (Figure 9.19).

The reinforcement should be laid on the blocks and then covered with a scoop of mortar. The next course of blocks should be laid while the mortar is still wet.

Reinforcement should be overlapped by a minimum of 100 mm where required.

Remember

Only bed-joint reinforcement that has been specifically manufactured for use with thin layer mortar masonry should be used with the Celcon thin joint system.

Openings

When forming window openings, the blockwork can be taken up in full-block heights and then cut blocks used to build up to sill level, lintel level and wall plate level.

Lintels

Several types of lintel can be used with thin joint systems, including steel box and combined lintels (Figure 9.20).

For cavity wall construction the box lintel is recommended. If a combined lintel is planned then the lintel will require propping until the outer leaf has been built to support the lintel and prevent it rotating.

For openings in solid walls the combined lintel is recommended.

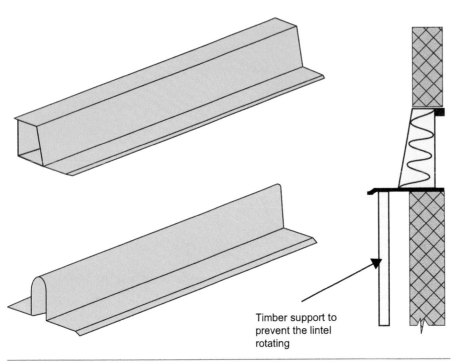

Timber support to
prevent the lintel
rotating

FIGURE 9.20
Lintels for cavity walls

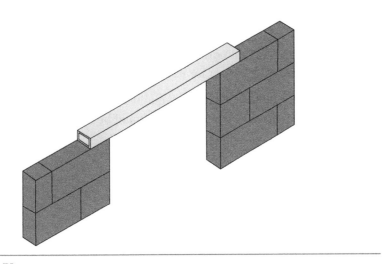

FIGURE 9.21
Lintels for internal walls

Openings in internal walls can be bridged with various types of lintel (Figure 9.21).

Damp-proof courses

These are positioned as for traditional blockwork.

Multiple-choice questions

Self-assessment

This section of the book is designed to allow you to check your level of knowledge. The section consists of revision questions for this chapter. The questions are all multiple choice and have four possible answers. The answers are to be found at the end of the book.

The main type of multiple-choice question will be the four-option multiple-choice question. This will consist of a question or statement, known as the stem, followed by a choice of four different answers, called the responses. Only one of these responses is the correct answer; the others are incorrect and are known as distracters.

You should attempt to answer the questions by choosing either (a), (b), (c) or (d).

Example

The person employed by the local authority to ensure that the Building Regulations are observed is called the:

    (a) clerk of works

    (b) building control officer

    (c) council inspector

    (d) safety officer

The correct answer is the building control officer, and therefore (b) would be the correct response.

Thin joint masonry

Question 1 Identify the following item used with thin joint masonry:

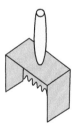

    (a) trowel

    (b) scoop

    (c) template

    (d) box square

Question 2 How is the thin joint mortar mixed?

    (a) by machine in a cement mixer

    (b) by hand with a shovel

    (c) by hand in a bucket

    (d) delivered to site ready mixed

Question 3 Which of the following bonds is used for thin joint masonry?

(a) English bond

(b) Flemish bond

(c) Dutch bond

(d) stretcher bond

Question 4 Which of the following is the recommended bed joint thickness?

(a) 2 mm

(b) 4 mm

(c) 6 mm

(d) 10 mm

Question 5 Identify the following wall tie:

(a) helical tie

(b) spiral tie

(c) twisted tie

(d) steel tie

Question 6 Compared with traditional blockwork, what is the main benefit of thin joint masonry?

(a) cheaper to build

(b) easier to build

(c) slower to build

(d) faster to build

Question 7 Where should a shear tie be used?

(a) at a junction with another wall

(b) at a corner

(c) across a movement joint

(d) across the cavity

Question 8 How are wall ties fixed to tie brick outer leaves to inner leaves in thin joint masonry?

(a) placed on a corresponding course

(b) driven into the thin block wall

(c) screwed to the thin block wall

(d) stuck to the thin block wall

CHAPTER 10

Bridging Openings with Lintels and Arches

This chapter will cover the following City and Guilds units:

- Apprenticeship   Module 7   S3.1
                                      K3.1
- Certificate   Unit 204   1.1, 1.2
                                      2.1, 2.2, 2.3, 2.4
- Diploma   Unit 206   6.1, 6.2, 6.3, 6.4, 6.5, 6.6, 6.7, 6.8, 6.9, 6.10

This chapter is about:

- Interpreting instructions
- Adopting safe and healthy working practices
- Selecting materials, components and equipment
- Preparing and erecting bridging to openings

The following Skills and Knowledge requirements will be covered:

- Set out and build a small rough ringed arch
- Know methods used to set out semi-circular and segmental arches

The following Certificate learning outcomes will be covered:

- Know how to set out semi-circular and segmental arches
- Know how to construct semi-circular and segmental arches

The following Diploma learning outcomes will be covered:

- Be able to form and bridge openings in cavity walls
- Know how to form semi-circular, segmental and rough ring arches

Building information

Whenever openings are formed some form of structure is needed to bridge the opening. This chapter deals with bridging openings with lintels and arches.

It is important that the student can understand drawings and schedules to be able to understand which method is used and where.

A typical lintel schedule is shown in Table 10.1. This explains to the builder what type of lintel is required and where it is fitted, and allows the builder to preorder the correct lintels for the works.

If the openings are to be bridged with arches the builder will require special arch centres for the bricklayers to construct the arches as the work proceeds.

Description and brief history

All buildings have openings of some description to allow access and egress and admit light and ventilation. Whenever an opening is made in brick or block-work walls a weakness could occur in the walls if the work is not carried out correctly.

Today's modern buildings include larger openings than before so it is essential to take extra care when forming the opening. The most appropriate method of bridging the opening must be selected to maintain the maximum amount of stability and strength in the wall.

These openings have to be bridged to support the walling above in a way that will not cause any cracking to the structure either above or below the opening. There are several methods available, the simplest of which is to use a lintel. The lintel carries the weight of the load above the opening and distributes this load to the abutments either side.

Early civilizations tried various ways of supporting brick and stone masonry across openings in their buildings (Figures 10.1 and 10.2). The pyramid-

Table 10.1 Lintel schedule

Job Number......   Site Address.........................

Location	Position and size	Type Steel boxed lintel	Precast concrete lintel	Corrugated steel
Dining	External 1.500	✓		
	Internal 1.100			✓
Lounge	External 2.000	✓		
	Internal 1.100			✓
Kitchen	External 1.500	✓		
	Internal 1.100			✓
Study	External 2.000	✓		
	Internal 1.100			✓
Garage	External 1.500	✓		
	Internal 1.100		✓	

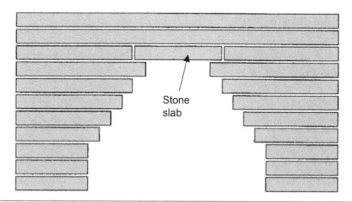

FIGURE 10.1
Spanning an opening by corbelling

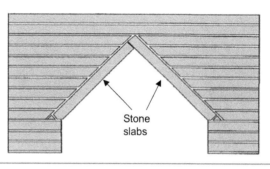

FIGURE 10.2
Alternative early method of spanning an opening

building Egyptians and ancient Greeks used lots of columns and stone lintels, but it was the later Roman builders who developed the idea of forming arches from separate blocks of stone, bricks or tiles mortared together.

Properly bonded brickwork can bridge across openings in a series of offsets (Figure 10.3). This is called natural bracketing and means that in theory, only the area of brickwork within the 60 degree triangular shape in Figure 10.3

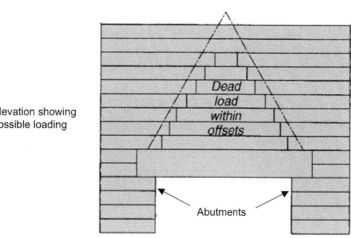

FIGURE 10.3
Natural bracketing of bonded brickwork above a lintel

needs supporting by a lintel. The greater the span of an opening, the larger this triangular piece will be.

Lintels

Definition

Lintels are straight beams of concrete or steel that are made strong enough to support the dead load of bricks or blocks above. Because lintels are straight, they have a slight tendency to bend or deflect when bricks or blocks are bedded on top. This loading results in tensile stress in the lower part of a lintel, plus compressive stress in the upper part (Figure 10.4).

Steel rods are cast into lower portion of concrete lintels to absorb this tension, so that the lintel does not crack on the underside when loaded. Concrete in the upper part of the lintel is very good at resisting compressive stress.

Reinforced concrete lintels

Concrete lintels can be made in any one of five ways, depending on the following site considerations:

- How large do the lintels need to be?
- How much will each weigh (mass in kg)?
- How are they to be moved around site and to upper floors?
- Are they to be lifted into place by hand or mechanically?
- Will they be seen in the finished work, or plastered?

Precast lintels

These lintels are made in a mould at ground level. When the concrete has hardened for at least a week, they are lifted and bedded in place.

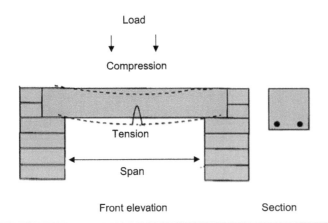

FIGURE 10.4
Typical stresses in a lintel

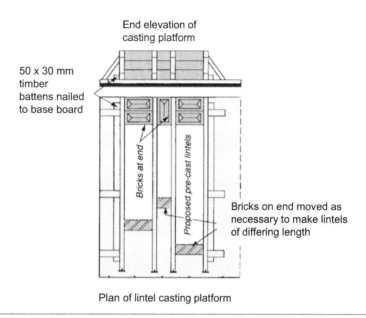

End elevation of casting platform

50 x 30 mm timber battens nailed to base board

Bricks at end

Proposed pre-cast lintels

Bricks on end moved as necessary to make lintels of differing length

Plan of lintel casting platform

FIGURE 10.5
Method of casting lintels on site

Figure 10.5 shows a typical site casting platform, set perfectly level both ways. Bricks or blocks are used as spacers, to obtain lintels of exactly the right width.

The inside surfaces of these lintel boxes should be lightly coated with mould oil to make dismantling or 'striking' easier. When filling the mould a 25–30 mm layer of wet concrete should first be spread over the bottom before placing reinforcing steel bars.

The remainder of the concrete should be added and thoroughly compacted by manual or mechanical vibration.

The upper surfaces should be trowelled smooth and marked clearly as the 'Top' surface, so that the steel bars will be towards the soffit when permanently bedded (Figure 10.6).

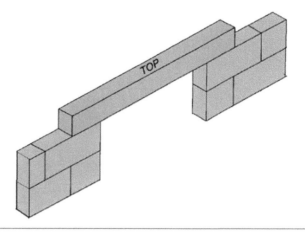

FIGURE 10.6
Precast reinforced concrete lintel in position

Factory-made reinforced concrete lintels

For larger requirements precast reinforced concrete lintels can be ordered from a specialist supplier to suit the range of openings in a building. Care should be taken that a fair face is specified where lintels will be permanently exposed, and that the top surface of all lintels delivered to site is clearly marked (Figure 10.6).

Prestressed concrete plank lintels

These special reinforced concrete lintels can only be made in a factory.

The high-tensile steel strands are stretched by hydraulic jacks before the concrete is compacted in the lintel mould. When the concrete has thoroughly hardened, the stretched steel reinforcement is carefully released and keeps the whole lintel section in a permanent state of compression. These lintels are known as plank lintels because they are only 65 mm deep for all spans up to 1800 mm (Figure 10.7).

Cast in situ lintels

These are used in cases where a lintel would be much too heavy to lift by hand if cast at ground level. Instead, a timber box or mould is constructed exactly where the lintel is required. This formwork (Figure 10.8) is made strong enough to withstand the pressures of filling with wet concrete and thoroughly compacting it.

The concrete is poured around steel reinforcing bars placed approximately 25 mm up from the soffit (Figure 10.9).

Bedding lintels

Lintels must be set upon a wet mortar bed joint at each side, so as to spread the load evenly over the whole bearing surface. When tapping down into position the spirit level should be held against the underside of the lintel in case the top edge is not parallel with the soffit.

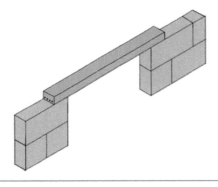

FIGURE 10.7
Prestressed concrete plank lintel in position

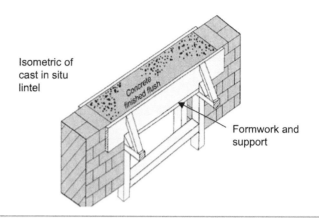

Isometric of cast in situ lintel

Formwork and support

FIGURE 10.8
Formwork for a cast in situ lintel

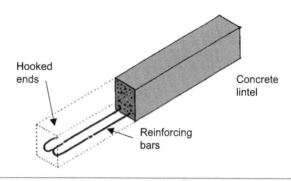

Hooked ends

Concrete lintel

Reinforcing bars

FIGURE 10.9
Reinforcement in lower part of lintel

Steel lintels

Sheet steel can be pressed into shape in a factory to make lintels that are much lighter to lift than concrete, and which can withstand both the tensile and compressive stresses.

These are available in several shapes and sizes.

Internal steel lintels

Internal steel lintels are designed mainly for 100 mm walls but other sizes are available. They can be boxed, corrugated or channelled (Figures 10.10–10.12).

External steel lintels

External steel lintels are designed mainly for cavity walls and are available in various designs (Figures 10.13–10.15).

Lintel bearings

A solid seating or bearing is required at each end of a lintel, to support the concentrated pressure of one half of the total load resting on the lintel.

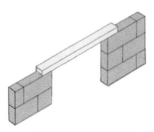

FIGURE 10.10
Boxed steel internal lintel

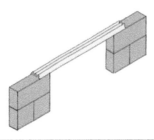

FIGURE 10.11
Corrugated steel internal lintel

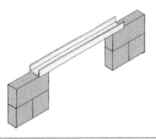

FIGURE 10.12
Channelled steel internal lintel

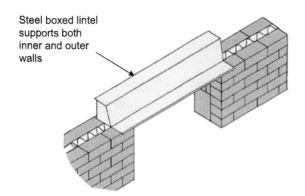

Steel boxed lintel
supports both
inner and outer
walls

FIGURE 10.13
Steel section lintel for cavity walls

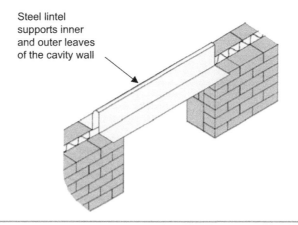

FIGURE 10.14
Bent steel lintel for cavity walls

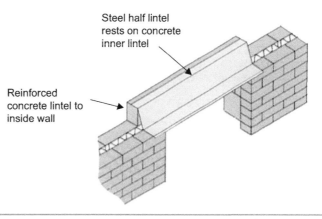

FIGURE 10.15
Half steel lintel for cavity walls

This bearing at each side should be a minimum of 150 mm for lintel spans up to 1800 mm, with 225 mm minimum up to a 3 m span. A structural engineer will be required to calculate the safe bearing area needed for larger span lintels. BS5628:1985 Part 3, 'Workmanship', recommends that this bearing level should coincide with a whole block rather than a half or cut block, for greater stability where lintels are set in block walling. See Figure 10.16 and all previous drawings.

Reinforced brick lintel

The bricklayer calls this a soldier arch, as the bricks are placed on end, resembling a file of soldiers.

Care should be taken to avoid laying the bricks at an angle. A small boat level should be used to check the work, as explained in Chapter 6 for soldier courses (Figure 10.17).

Before setting the lintel, temporary supports are placed (Figure 10.18). Several methods of reinforcing brick lintels are shown in Figure 10.18: at 'A', where a 225 mm soffit occurs, steel bar reinforcement is placed throughout the length of the

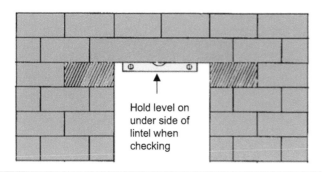

FIGURE 10.16
Whole block under lintel bearings to spread the load

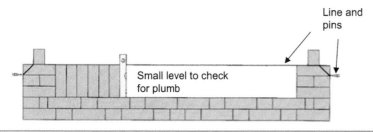

FIGURE 10.17
Checking soldier course bricks

lintel, with stainless steel ties positioned at intervals of approximately 225 mm; at 'B' the brick lintel is erected with wire reinforcement built in and left projecting, to allow the concrete lintel to be cast in situ around it, thus giving adequate stability.

This type of lintel, if erected truly horizontal and plumb, causes an optical illusion to occur at a distance of 3 m or more from it; it appears to sag. This appearance can be overcome by giving the lintel a slight camber or by giving a curvature to the soffit along its length, together with a slight skewback. The adjustment must be made carefully and with the consent of the architect.

Brick arches

Arch construction is a more decorative means of spanning openings. True arches are made to curve upwards when looked at in elevation, so that they are always in a state of compression, wedged between abutments. As bending under load cannot take place, there will be no tensile stress in the arch.

All arches are formed on a temporary support called an arch centre, which must not be removed until the jointing mortar has been allowed to harden for at least a week. Only when the arch centre has been removed does the arch become self-supporting.

Terminology

Arches comprise small units bonded together around a curve.

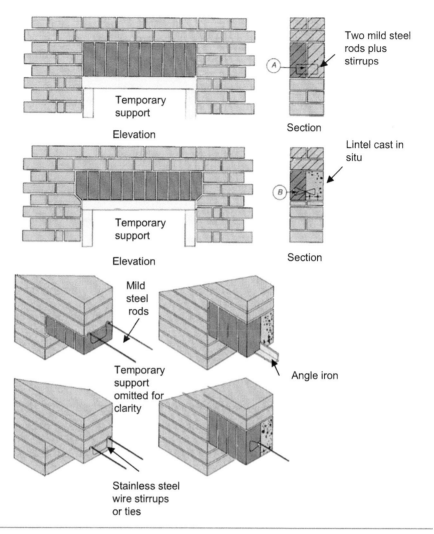

FIGURE 10.18
Soldier arches and reinforced lintels

If an arch is constructed correctly it will not require any additional reinforcement, as brick lintels do, because the units are wedge shaped and any load applied on to the arch will tighten the units together.

The curved shape of an arch distributes the load of the walling above an opening down through the abutments on either side of the opening.

Arches are classified generally into three groups according to their shape, the number of centres and the method of cutting and preparing the arch.

The names of the parts of an arch are shown in Figure 10.19.

Arch types

The three types of arch are:

- rough arches
- axed arches
- gauged arches.

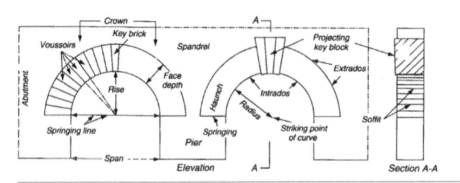

FIGURE 10.19
Craft terms in arch construction

Rough ringed arch

This type of arch is the simplest to construct. It is formed of uncut bricks and its shape is controlled by the type of turning piece or centre adopted.

The joints are wedge shaped and thus play an important part in the stability of the arch.

Their use is usually restricted to semi-circular and segmental arches (Figure 10.20).

The arch is constructed of a number of half-brick rings, hence the name ringed arch, the number of half-bricks varying according to the size of the opening.

This method is adopted to reduce the size of the mortar joint at the extrados of each separate ring.

Axed arches

This type of arch is formed by bricks cut to a wedge shape using an ordinary facing brick (Figure 10.21). It can be of the same colour and texture as the wall facings, or of some contrasting colour and probably of finer texture.

The tools used for cutting and trimming are the hammer, bolster and scutch, with a carborundum block.

(a)   (b)

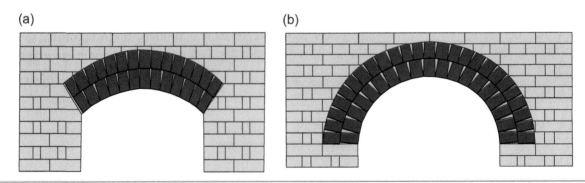

FIGURE 10.20
Rough ring arches: (a) segmental arch; (b) semi-circular arch

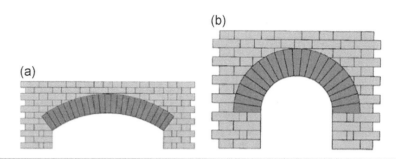

FIGURE 10.21
Axed arches: (a) segmental arch; (b) semi-circular arch

The thickness of the mortar joint varies from 4 to 10 mm.

The operations of cutting and setting will be explained later in this chapter. It is also possible to bond the face of an axed arch.

Gauged arches

These arches are sometimes known as rubbed arches and have a much finer joint between the bricks. It is usually 2 mm and made from lime putty (Figure 10.22).

This work requires specially made clay bricks called red rubbers. These bricks contain approximately 30 per cent fine sand mixed with the clay, and are carefully fired so that the same even orange–red colour and texture is present throughout the body of each brick.

The high content of sand allows these bricks to be hand sawn and rubbed, on a block of York stone, to permit joints of 2 mm thickness.

Gauged work demands the very highest level of skills from the bricklayer in setting out, cutting, rubbing and bedding these bricks, which are 'white-line' jointed, using a putty made only from freshly slaked lime and water.

Semi-circular arches

As the name implies the arch forms a semi-circle. Any load placed on the walling above the arch is transferred vertically down through the arch onto the abutments.

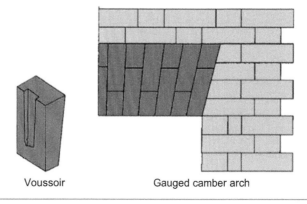

Voussoir   Gauged camber arch

FIGURE 10.22
Gauged arch

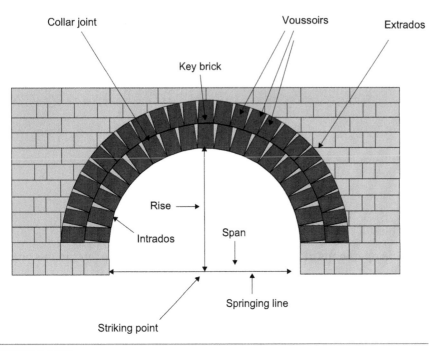

FIGURE 10.23
Semi-circular arch details

Figure 10.23 shows the various parts of a semi-circular arch.

Geometry of semi-circular arches
Information required:

- the span – assume 150 mm (using a scale of 1:10)

- depth of face – assume one brick in two rings.

Construct a 150 mm span and bisect to find the centre (Figure 10.24).

Place a compass point on the centre point and set the radius to half span. Scribe the arch. This curve is the intrados of the arch.

The depth can be set out, say 22.5 mm, and the extrados drawn.

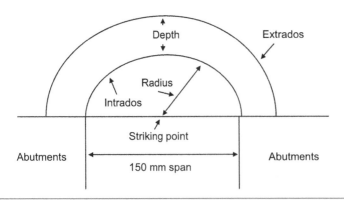

FIGURE 10.24
Geometry of a semi-circular arch

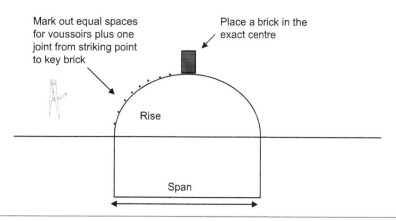

Mark out equal spaces for voussoirs plus one joint from striking point to key brick

Place a brick in the exact centre

Rise

Span

FIGURE 10.25
Setting out the key brick and the number of voussoirs

The centre brick, known as the key brick, is marked out first. When marking out in situ, a brick can be placed in the required position at the top of the arch, allowing space for a joint either side (Figure 10.25).

Use dividers to set out the voussoirs of the arch from the key bricks back down to the springing line either side.

The dividers should be set to the width of a brick plus a 10 mm joint. This should always give an odd number of voussoirs.

When drawing the voussoirs the key brick should be marked out first in each ring.

Draw a circle equal to the width of the brick around the striking point.

A line is then drawn from either side of the circle through the marked out voussoirs to give the wedge-shaped joint (Figure 10.26).

Constructing a semi-circular arch

The method of constructing a semi-circular arch is similar for all types (Figure 10.27).

Set out the opening with a tolerance of 5 mm.

Use the arch centre as a guide or produce a pinch rod, to check the abutments for plumb as the work proceeds.

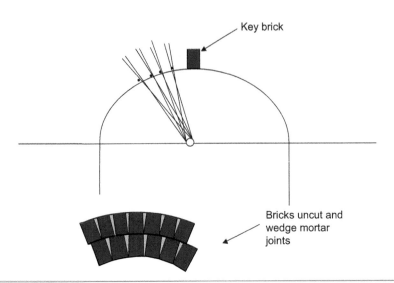

FIGURE 10.26
Marking out the wedge-shaped joints

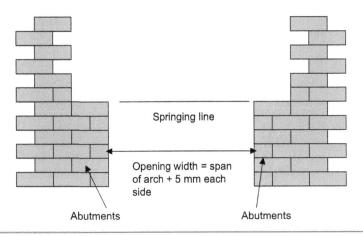

FIGURE 10.27
Building up the abutments to receive the arch centre

Build the abutments up to the springing line. Position the centre supported on timber props and folding wedges. These will be dealt with later in this chapter.

It is advisable to build the abutments to the height of the crown of the arch as support for the lines.

Position and fix the arch centre using folding wedges for easy removal after the arch has been built.

Set dividers to the thickness of a brick plus a normal joint and step out around the arch centre, not forgetting to have a key brick in the centre (Figure 10.28).

Pull a line through from the corners and check that the centre is correctly placed and not interfering with the line.

Lay a few voussoirs on either side of the arch; never build one side completely.

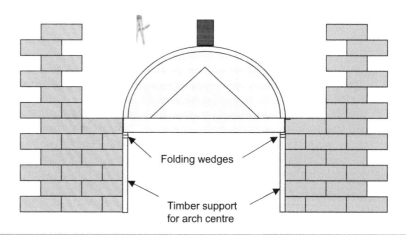

FIGURE 10.28
Setting the arch centre and marking out the voussoirs

It may be necessary to raise the line while the voussoirs are being laid.

Lay a second layer, if required, as before, keeping the key brick in the centre.

Raise the line course by course to complete the cutting around the extrados of the arch.

Always ensure that the joints are full to provide maximum strength to the arch. Strike the centre as soon as the arch is complete to allow the joints to contract uniformly when setting (Figure 10.29).

Always remember to joint the soffit of the arch immediately the centre is dropped.

Segmental arches

As the name implies, a segmental arch forms part of a circle.

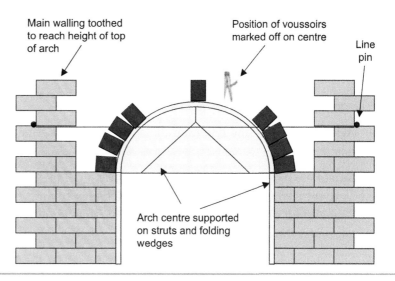

FIGURE 10.29
Completing the arch

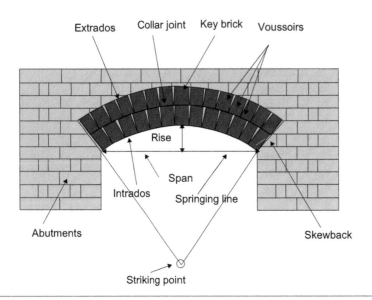

FIGURE 10.30
A two-ring, rough, segmental arch

Any load placed on the walling above the arch is transferred vertically down through the arch onto the abutments.

Figure 10.30 shows the various parts of a segmental arch. Recap on previous information if necessary.

Geometry of segmental arches

Again, the span must be known, but in the case of the segmental arch we also need to know the rise.

The rise to any segmental arch is normally one-sixth of the span.

First draw the span – assume it to be 900 mm – and set up a perpendicular bisector.

Mark off the rise: 150 mm. Join AB and bisect.

The point where this bisector intersects the centre line, C, is the striking point of the required arch.

A face depth of 225 mm is shown in Figure 10.31.

The angle produced by the line from the striking point, passing through the springing point to the extrados, is known as the skewback.

To ascertain the number of voussoirs and the position of the joint lines in the axed arch, set the dividers at the width of a brick plus 10 mm, and place the points equidistant on each side of the centre line, thus forming the key brick and step round the intrados.

If the last step fails to connect with the springing point at the first attempt, make a further attempt by slightly closing the dividers.

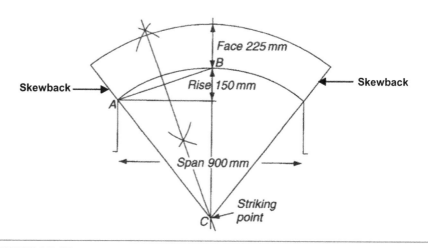

FIGURE 10.31
Geometric setting out of segmental arch

This procedure is exactly the same as for the semi-circular arch. The angle of skewback can now be set using one of two methods.

Place a brick on the centre adjacent to the springing point and set a sliding bevel to the angle produced.

Pull a line from the striking point through the springing line at the point of support and set the bevel to this angle (Figure 10.32).

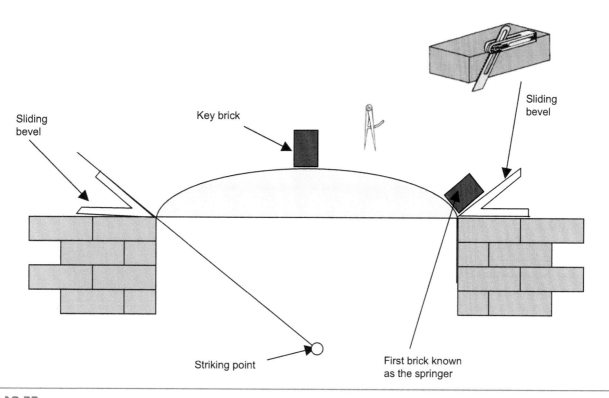

FIGURE 10.32
Setting out skewback

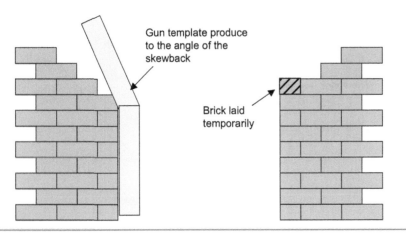

FIGURE 10.33
Use of the gun template

Cutting skewbacks

For cutting the skewbacks in preparation for the building of a segmental arch, always use a gun template.

Its function is to keep the angle of the skewback constant, especially when there are several arches to be constructed (Figure 10.33).

Constructing a segmental arch

The method of constructing a segmental arch is similar for all types.

Set out the opening with a tolerance of 5 mm.

Use the arch centre as a guide or produce a pinch rod, to check the abutments for plumb as the work proceeds.

Build the wall up to the springing line, and position the centre supported on timber props and folding wedges. Extend the abutments to a height above the top of the arch to support the lines (Figure 10.34).

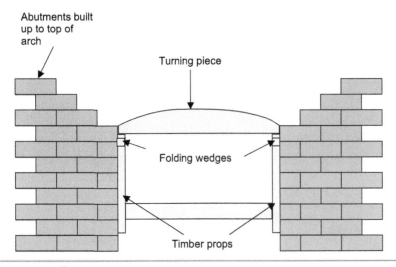

FIGURE 10.34
Setting the turning piece for a segmental arch

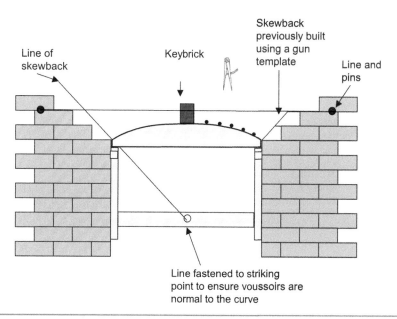

FIGURE 10.35
Setting the skewbacks and voussoirs for a segmental arch

Skewbacks must be marked and cut (Figure 10.35).

Pull a line from the striking point through the springing line at the point of support and set the bevel to this angle (Figure 10.32) or use a gun template.

Set dividers to the thickness of a brick plus half a normal joint and step out around the arch centre, not forgetting to have a key brick in the centre.

Pull a line through from the corners and lay the voussoirs to the line.

Lay second layer if required as before, keeping the key brick in the centre. Raise the line course by course to complete the cutting around the extrados of the arch. Strike the centre as soon as the arch is complete to allow the joints to contract uniformly when setting.

Always remember to joint the soffit of the arch.

Relieving or discharging arch

This is a form of ringed arch, built over a lintel, and turned on a shaped brick core (Figure 10.36). It relieves the lintel of any point load and discharges the weight to the abutments.

The joint between the brick core and the soffit of the arch must be of sand, as this ensures the proper discharging of loads. This sand joint is pointed on the face only.

The shape of the brick core is obtained using a template.

Types of support

No attempt will be made here to describe temporary arch supports over large spans. Only those encountered by the bricklayer in the course of everyday work will be dealt with.

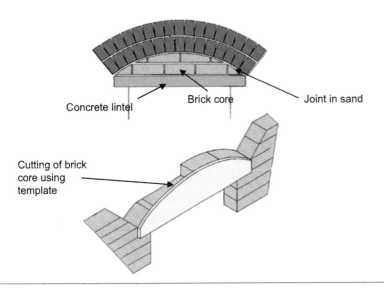

FIGURE 10.36
Formation of relieving or discharging arch

Turning piece

A turning piece is made from solid timber for arches of limited span.

For an arch with a 225 mm soffit, two 75 mm timbers can be placed side by side. Alternatively, a turning piece could be constructed from plywood and packed out with timbers. See examples in Figure 10.37.

Arch centres

Arch centres are usually produced for larger spanning arches. They can either be open or closed lagged:

- An open-lagged centre is framed from light timber and used for the turning of ringed arches (Figure 10.38).

- A close-lagged centre is suitable for the turning of an axed or gauged arch. Its use facilitates the marking of the voussoir positions (Figure 10.39).

When using temporary arch supports, folding wedges are placed directly under the centre (Figure 10.40). This facilitates the removal of the centre or turning piece and, in particular, avoids the chipping of the arch face on the intrados arris, which invariably occurs if wedges are not used.

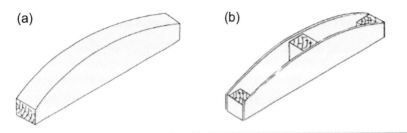

FIGURE 10.37
Arch turning pieces: (a) solid turning piece; (b) built-up turning piece

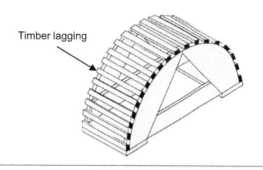

FIGURE 10.38
Open lagged centre for semi-circular arch

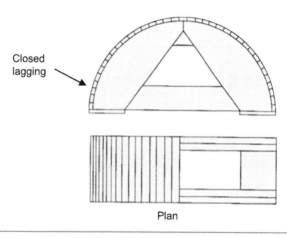

FIGURE 10.39
Closed lagged centre for semi-circular arch

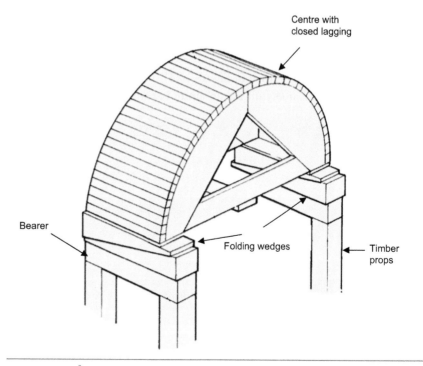

FIGURE 10.40
Centre supported on timber props and folding wedges

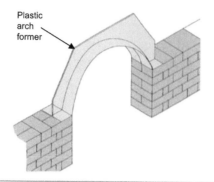

FIGURE 10.41
Plastic arch former for solid walls

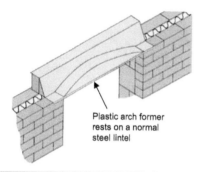

FIGURE 10.42
Plastic arch former sitting on a boxed steel lintel on cavity walls

Modern methods

There are also new methods of supporting arches in both solid and cavity walls. These arch formers are available in plastic and steel.

The centres are built in and remain as part of the structure.

A semi-circular centre is shown in Figure 10.41, but segmental centres are also available. A version that sits on a steel lintel is shown in Figure 10.42.

Multiple-choice questions

Self-assessment

This section of the book is designed to allow you to check your level of knowledge. The section consists of revision questions for this chapter. The questions are all multiple choice and have four possible answers. The answers are to be found at the end of the book.

The main type of multiple-choice question will be the four-option multiple-choice question. This will consist of a question or statement, known as the stem, followed by a choice of four different answers, called the responses. Only one of these responses is the correct answer; the others are incorrect and are known as distracters.

You should attempt to answer the questions by choosing either (a), (b), (c) or (d).

Example

The person employed by the local authority to ensure that the Building Regulations are observed is called the:

    (a) clerk of works

    (b) building control officer

    (c) council inspector

    (d) safety officer

The correct answer is the building control officer, and therefore (b) would be the correct response.

Bridging openings

Question 1 What are the bricks laid vertical over openings known as?

    (a) soldier arch

    (b) vertical arch

    (c) flat arch

    (d) brick arch

Question 2 Which of the following is the correct name for the bricks in an arch?

    (a) key bricks

    (b) voussoirs

    (c) soldiers

    (d) soffit bricks

Question 3 What is the timber support to a semi-circular arch known as?

    (a) formwork

    (b) springer

    (c) centre

    (d) turning piece

Question 4 What is the timber support to a segmental arch known as?

(a) formwork

(b) springer

(c) centre

(d) turning piece

Question 5 Which of the following items of equipment can be used for setting out arch curves?

(a) gauge

(b) turning piece

(c) trammel

(d) datum

Question 6 What is the name given to the slope of brickwork cut to receive an arch?

(a) skewback

(b) springer

(c) closer

(d) voussoir

Question 7 An arch built with wedge-shaped joints is known as a:

(a) cut arch

(b) rough arch

(c) axed arch

(d) soldier arch

Question 8 Where should reinforcement be placed in a concrete lintel?

(a) at the top

(b) at the bottom

(c) just up from the bottom

(d) just below the surface

CHAPTER 11

Domestic Drainage

This chapter will cover the following City and Guilds units:

- Apprenticeship   Module 6   S1.1, S1.2
                                     K2.1, K2.1, K2.3, K2.4
                   Module 13   S1.1, S1.2, S1.3, S1.4
- Certificate   No comparable unit
- Diploma   No comparable unit

This chapter is about:

- Interpreting instructions
- Adopting safe and healthy working practices
- Selecting materials, components and equipment
- Preparing, laying and testing foul and surface drainage
- Construct an inspection chamber

The following Skills and Knowledge requirements will be covered:

- Construct an inspection chamber incorporating a half channel and branches
- Fit an inspection cover
- Understand how domestic drainage works

The following Certificate learning outcomes will be covered:

- No comparable unit

The following Diploma learning outcomes will be covered:

- No comparable unit

Interpreting technical information

There are numerous sources of information available and they include the following: drawings, programmes of work, schedules, specifications, policies, mission statements, manufacturers' technical information, organizational documentation, and training and development records and documents.

It is important that the student understands where information can be found.

Information sources

Information can be gathered from numerous sources and can be either written, oral or in drawing format.

Written

Manufacturers' technical information

Any efficient office should have up-to-date information regarding new and existing products from the various manufacturers who produce materials of interest to them.

Many manufacturers produce technical information which is free for the asking. These papers, brochures, leaflets, etc., should be stored in the office and sent to site as required.

Oral

Verbal communication can include talking on the telephone, mobile or walkie-talkie. This form of communication is instant and feedback can be obtained immediately.

Drawings

Drawings are one of the simplest forms of communication available. They are used extensively in the construction industry for transmitting ideas and information.

The types of drawing used range from simple sketches, to scale plans, graphs, product range drawings and charts. The type of drawing used will for the most part depend on the information to be transmitted and the time it is to be used.

Simple sketches can be used on site to show how a detail is to be built at that time, but if a few days or weeks are to pass before it is to be built, then a scale drawing will be more useful.

Drawings have been discussed in detail in Chapter 3. This chapter will only deal with drawings relating to drainage.

Understanding drawings

Everyone in construction needs to be able understand and extract information from drawings.

Table 11.1 Preferred scales for building drawings	
Type of drawing	Scales
Block plans	1:2500, 1:1250
Site plans	1:500, 1:200
Location drawings	1:200, 1:100, 1:50
Range drawings	1:100, 1:50, 1:20
Detail drawings	1:10, 1:5, 1:1
Assembly drawings	1:20, 1:10, 1:5

There are many types of construction drawings, which are listed in Table 11.1 along with the appropriate scales.

Production drawings

These are known as working drawings and are used to relay information to the building contractor and other members of the building team.

The working drawings can be classified as:

- location drawings
- component drawings
- assembly drawings.

Location drawings

These are further classified into:

- block plans
- site plans
- location plans.

Block plans

These are used to identify the site in relation to the surrounding area (Figure 11.1).

The scale is usually 1:2500 or 1:250 and is too small to allow much more than an outline of the site and boundaries, road layouts and the other buildings in the near vicinity.

The orientation of the site is always shown with a suitable logo depicting north. The actual site should be outlined in red.

It is unlikely that dimensions would be added to these drawings.

Plans of this sort are usually based on the Ordnance Survey sheet for the area; however, if such a source is used, permission should be obtained for its reproduction.

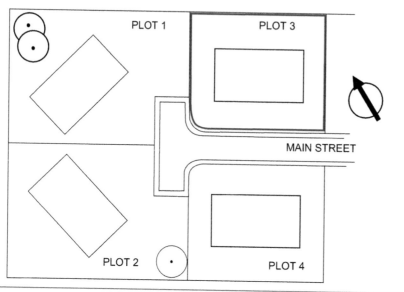

FIGURE 11.1
Block plan

Site plans

These are used to show the position of the proposed building on the site, together with information on proposed road, drainage and service layouts, and other site information such as levels. A typical site plan is shown in Figure 11.2.

Again, the orientation of the site should be shown.

Scales of 1:500 and 1:200 are often used.

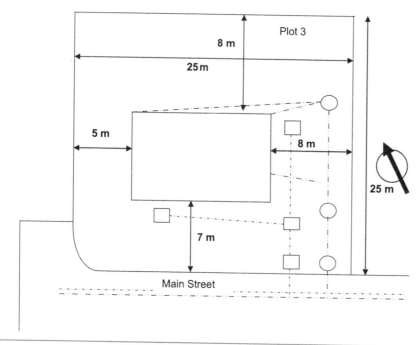

FIGURE 11.2
Site plan

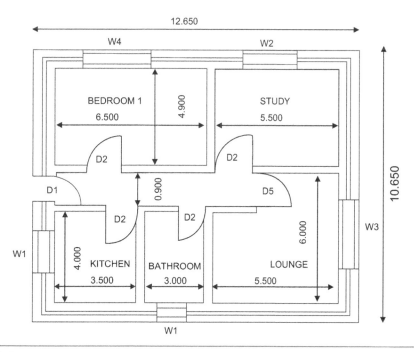

FIGURE 11.3
Location plan

The information on this drawing is used by both the design team and the contractors.

The drainage layout should be used in conjunction with the drainage schedule which will give details of inspection chamber construction, cover and invert levels, and other relevant information.

Location plans

These are used to show the size and position of the various rooms within the buildings and to position the principal elements and components. A typical location plan is shown in Figure 11.3.

Plans are usually drawn to a scale of 1:200, 1:100 or 1:50.

These scales are sufficiently large to allow dimensions to be added.

Further information from the drawing includes the wall construction and position and type of doors and windows. This information is used in conjunction with door and window and drainage schedules. The way the doors are hung is also shown.

Component drawings

These are further classified as range and details drawings.

Range drawings

These show the basic sizes and reference system of standard components.

They are usually drawn to a scale of 1:100, 1:50 or 1:20.

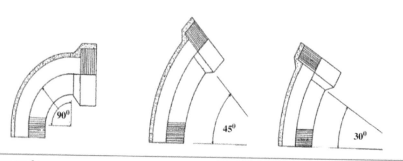

FIGURE 11.4
Range of bends

The types of bends can be obtained from any drainage manufacturer's litera-ture with their references, e.g. quarter bend, eighth bend or twelfth bend. This reference is all that is required by the architect when completing the working drawings, as the builder is provided with a drainage schedule for cross-referencing (Figure 11.4).

Assembly drawings

These are used by the architect to show in detail the junction between the various elements and components of the buildings (Figure 11.5).

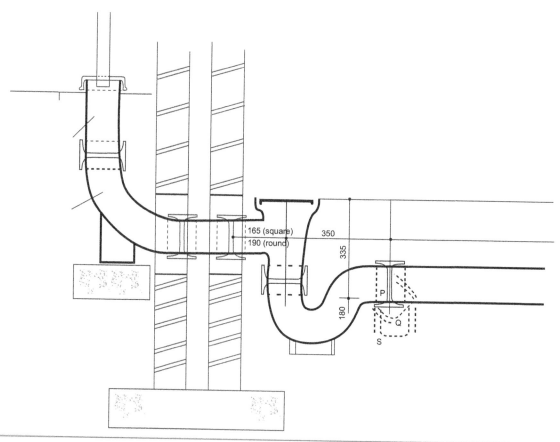

FIGURE 11.5
Assembly drawings

These details are necessary for the builder to know exactly how the architect requires the construction to be completed.

The scales are usually 1:20, 1:10 and 1:5 and should be fully dimensioned and annotated.

Schedules

Schedules form an important method of communication for the design team. These are prepared for large contracts to simplify tabulated information about a range of components.

The main areas where schedules are used include:

- doors
- windows
- ironmongery
- internal finishes
- sanitary ware
- drainage
- lintels
- steel reinforcement.

Some schedules are best produced by the specialist; for example, the steel reinforcement schedule by the structural engineer and the sanitary ware schedule by the mechanical engineer.

The information received from the schedule enables the quantity surveyor to prepare the bills of quantities more easily and more accurately.

By using the schedule, the main contractor and subcontractors can carry out certain operations more quickly, without the need to communicate with the architect constantly.

The schedules are additional aids to the specifications, drawings and the bill of quantities, and are extremely useful when locating work and checking deliveries of materials and components.

Producing and reading schedules is fairly straightforward.

The manufacturer's detail of the component is used for reference and the plans of the building.

The design of a schedule can take various formats, but normally the component types are listed across the top with the various finishes, and the number of units is listed down the left-hand column.

In the inspection chamber schedule shown in Table 11.2, the inspection chambers (ICs) are listed down the left-hand column and the various designs and sizes are shown across the top.

Table 11.2	Inspection chamber schedule								
Inspection chamber	Size	Invert depth	Connections	Pipe size	Step iron	Cover	Internal finish	Comments
IC1	450 × 450	450	1	100	0	450 × 450	Brick pointed	
IC2	450 × 450	600	1	100	0	450 × 450	Brick pointed	
IC3	450 × 750	750	2	100 150	1	450 × 450	Brick pointed	
IC4	450 × 750	1000	2	150	2	600 × 600	Brick pointed	
IC5	750 × 1200	1200	3	150	2	600 × 600	Brick pointed	

The completed schedules will be used by the:

- quantity surveyor for cost planning and production of the bill of quantities

- structural engineer for reinforcement bending schedules, beam, column and foundation schedules

- services engineer for builder's work details, fire stopping, sanitary ware and luminaries

- builder as part of the production information used for tendering and contract control.

Taking off quantities

At some stage in your career you will have to determine the quantity of materials required to carry out an activity. This will entail having to calculate areas and volumes.

Quantities of materials

The buyer is responsible for taking off and scheduling materials from the bills of quantities or, on smaller jobs, the working drawings.

When the site supervisor if given this role he or she must be careful to ensure that correct quantities and details of quality of materials are extracted from the drawings and specifications.

Allowances should be made for materials wastage which, from experience, is usually 5–10 per cent.

Taking off materials measurements from drawings should be carried out with care to allow for extra lengths or quantities, especially on joists, to give a suitable margin of safety.

Too many examples are noticed on building sites of long offcuts from joists, rafters and other structural members owing to generous allowances having been made at the taking-off stage. Extra members can also be seen lying around the site unused.

Examples

Bricks

The format size for a standard brick (plus joint) is 225 mm × 112.5 mm × 75 mm.

This means that a square metre of a one-brick-thick wall requires 118.5 bricks, but allowing 5 per cent for cutting and waste a more realistic figure is 125 bricks.

To calculate the total number of bricks for a one-brick wall, calculate the area in square metres and multiply by 125.

Example 1

A one-brick wall is 6.00 m long and 1.50 m high. Calculate the number of bricks required.

Answer 1

$$6.00 \times 1.50 = 9.00 \text{ m}^2$$

$$9.00 \times 125 = 1125 \text{ bricks required}$$

Concrete

To calculate the volume of concrete foundations you need three dimensions: the length of the trench, the width of the trench and the thickness of the concrete foundations.

Example 2

Calculate the concrete required for a foundation if the trench is 39.00 m long and 0.75 mm wide and the concrete is 0.15 mm thick. This would be exactly the same for pipe bedding.

Answer 2

$$39.00 \times 0.75 \times 0.15 = 4.387 \text{ m}^3$$

Drainage

The number of pipes can be calculated by dividing the length of the trench by the length of each individual pipe.

Example 3

Calculate the number of drain pipes required for a trench 36.00 m long if 0.900 mm pipes are being used.

Answer 3

$$36.00 \div 0.900 = 40 \text{ pipes}$$

Gradients

The most common gradient is stated as 1:40. If we use a drain length of 40 m it will simplify the calculations.

Example 4

Calculate the fall of a drain trench 40 m long if the gradient has to be 1:40.

Answer 4

$$\frac{40 \times 1}{40} = 1 \text{ m}$$

Example 5

Calculate the fall of a drain trench 25 metres long if the gradient has to be 1:40.

Answer 5

$$\frac{25 \times 1}{40} = 0.625 \text{ mm}$$

Safe working practices

It is essential that the correct personal protective equipment (PPE) is available and worn by all operatives working on drainage schemes.

As most domestic drainage is below ground in trenches the correct care and attention must be taken. Two types of accidents can occur with excavations: either the trench itself collapses or people fall into the trench. Always make sure that the trench is well supported, keeping heavy loads away from the edges, and erect barriers around the excavation where necessary (Figure 11.6).

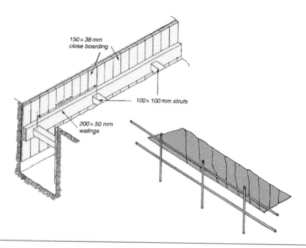

FIGURE 11.6

Timbering to a drain trench and barriers along a trench

FIGURE 11.7
Service posts

Accidents should only occur in unforeseen circumstances.

Anyone who sees a hazard and does nothing about it is making it much more likely that an accident will occur. It is essential to make safe or report dangerous situations.

When excavating trenches the normal attention should be paid to existing underground services. These can be located by contacting the local services and with the use of a cable locater. Look out for tell-tale signs and clues in roadside service signs such as those shown in Figure 11.7.

Material and component suitability

This chapter deals with various walling and drainage components and materials, including:

- standard and engineering quality bricks
- types of cement and mortar additives
- drainage
- inspection chamber covers and frames
- step irons
- plastic and concrete inspection chambers
- bricks.

(a)   (b)

(c)

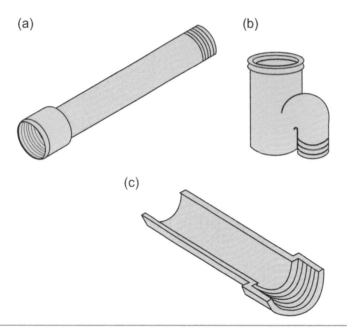

FIGURE 11.8
Selection of clayware: (a) pipe; (b) gully with 'S' trap; (c) channel

Drainage pipes

Drainage pipes can be manufactured from various materials and are classed as either rigid or flexible.

Clay is the main material used for rigid drain pipes in domestic work, with cast iron the main alternative. The usual materials for flexible drain pipes are pitch fibre and unplasticized PVC. A selection of clayware is shown in Figure 11.8.

Inspection chamber covers

These are produced from concrete, cast iron or steel. They come in various shapes and sizes to suit the particular situation. Figure 11.9 shows two examples.

Inspection chambers (manholes)

Inspection chambers are usually constructed in engineering bricks (Figure 11.10). Other methods on the market include plastic and precast concrete chambers.

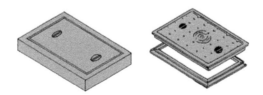

FIGURE 11.9
Concrete and steel covers and frames

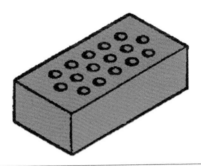

FIGURE 11.10
Engineering brick

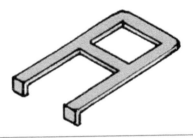

FIGURE 11.11
Galvanized steel step irons

When constructing deep inspection chambers in brick, step irons will be required (Figure 11.11).

Mortar for constructing inspection chambers

Cement and sand
These materials are used to produce mortars for construction brick inspection chambers. Cement should be ordinary Portland cement, or sulphur resisting if required, and sand should be clean building sand.

Plasticizers
Plasticizers are added to mortar mixes to increase the workability of the mix.

Accelerators
These are added to mortar or concrete mixes to speed up the initial setting time.

Selection of mortar

As a general guide, the hardness or eventual compressive strength of mortar should be the same as the hardness of, or preferably slightly weaker than, the bricks to be laid. Therefore, if, as a result of slight foundation settlement, cracks develop, these will follow the joint lines, which can easily be cut out and repointed.

Excessively hard mortar can result in the bricks becoming fractured at settlement cracks, leading to a more extensive repair operation. Another reason for relating strength of mortar to the compressive strength of the bricks is to ensure that external walls weather evenly during the lifetime of the building, with any absorbed water evaporating at a similar rate from the surfaces of bricks and joints alike.

In addition to being hard enough to transfer loads evenly between irregular surfaces of bricks, the mortar must resist the effects of rain and frost in the long term.

Drainage systems

The purpose of a drainage system is to convey waste products from toilets, bidets, baths, hand basins and sinks from all public and private buildings.

The waste must be conveyed in a manner that will not cause any unpleasant odours or hazards to health. It is conveyed through a series of impervious drain pipes laid to falls which take the contents to a place where it can be treated and disposed of.

All drainage work must be carried out in accordance with the current Building Regulations.

Definitions

Drain

This includes all pipes, fittings, gullies and inspection chambers laid to remove soil water, waste water or surface water from one building. It is the property of one owner, who is responsible for its maintenance and repair.

Private sewer

This includes all pipes, fittings, gullies and inspection chambers laid to remove soil water, waste water or surface water from *two or more* buildings. It is the joint property of two or more owners, who are responsible for its maintenance and repair.

Public sewers

This is any length of drain that is not covered by the definition of drain or private sewer and is the property of the local authority, which is responsible for its maintenance and repair.

Sewage

Effluent or sewage can be defined as unwanted water that has to be disposed of. It may comprise water-borne human, domestic, trade or farm waste, and it may also be surface water or subsurface water.

Types of water

There are two types of water in drainage systems:

- surface water
- foul water.

Surface water may be collected from the roofs of buildings and the paved areas around them. This water is collected in the roof gutters and in gullies from paved areas, and discharged directly into the drainage system.

Water which is collected from sanitary appliances such as sinks, baths and toilets is said to be foul water.

Once water has been introduced into a building it will need to be discharged, whether contaminated or not. Any system that carries this effluent within the building site is called a drainage system and is the responsibility of the owner, but once this system passes to the highway it becomes a sewage system and is the responsibility of the local authority.

Systems

There are two main systems currently in use, with a possible third in some areas:

- combined system
- separate system
- partially separate system.

This system conveys waste water from both surface water and foul water in a single pipe to the sewage treatment works. This is very common in certain parts of the country and is often used when treatment is not available, e.g. when waste is fed into the sea.

It is the simplest and cheapest system, with no chance of connecting to the wrong pipe.

Drains are well flushed during heavy rainfall and there is only one pipe to maintain. Silting of the pipes may occur in dry weather unless self-cleansing velocities are reached during foul flow – not during storms. At treatment works spasmodic rushes of water during storms can cause operational difficulties. A typical combined system is shown in Figure 11.12.

This system involves two separate pipes, one carrying the foul water to the sewage treatment works, and one carrying the surface water directly into a water course or the sea.

In house drainage it requires two sewer connections, with two pipes to maintain. Sometimes the pipe layout is difficult and costly and pipes may cross each other.

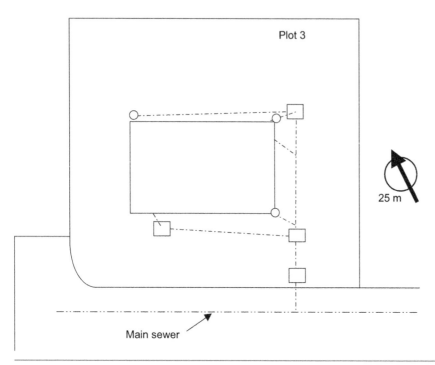

FIGURE 11.12
Combined system of drainage

The foul drain cannot be flushed by rain water, but surface water is well flushed in times of heavy rain. The treatment plant receives a flow of regular strength and sewers must be laid to correct gradients to give self-cleansing velocities. A typical separate system is shown in Figure 11.13.

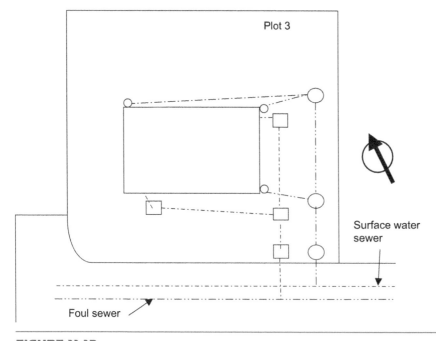

FIGURE 11.13
Separate system of drainage

Partially separate system

This is a combination of the two previous systems where a proportion of the surface water is conveyed to the surface water sewer, soakaway or watercourse, and the remainder is connected to the foul water system.

It is usual to connect the farthest rain water gully to the foul water system to assist in flushing out the system during heavy rain storms.

A typical partially separate system is shown in Figure 11.14.

Pipes

Pipes are manufactured from various materials, such as:

- Clay
- UPVC
- concrete
- cast iron.

Pipes and fittings must be sound and free from visible defects.

They are available in either salt glazed or ceramic glazed finishes.

All pipes manufactured in accordance with the requirements of the British Standard (BS) specification should be clearly marked, with the identifying mark of the institute stamped on them.

Clay pipes

Clay pipes are manufactured with or without a socket (Figure 11.15).

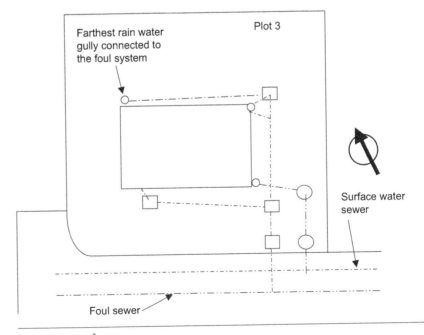

FIGURE 11.14
Partially separate system of drainage

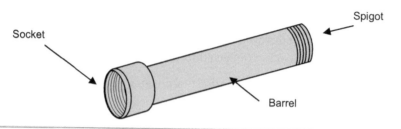

FIGURE 11.15
Clay pipe

Pipes must bear the manufacturer's name or trademark. They should also carry the BS kite mark.

Joints

All pipes can be connected with a spigot and socket joints. Some pipes are now produced with plain ends.

Rigid joints

Each pipe is laid to line and the joint is caulked with gaskin or tarred rope. This centralizes the pipe and prevent the cement mortar from entering the barrel of the pipe (Figure 11.16).

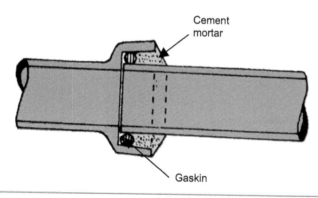

FIGURE 11.16
Rigid joint

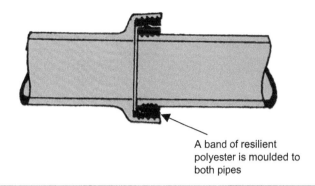

A band of resilient polyester is moulded to both pipes

FIGURE 11.17
Flexible joint

Flexible joints

There are various types of flexible joint.

In some cases the jointing material is fixed to the pipes during manufacture (Figure 11.17). Others have special couplings, which when greased, slide over the ends of pipes to form a watertight and flexible joint. The pipes in this method have plain ends (Figure 11.18).

Bedding

The code of practice for building drainage recommends the following bedding for rigid and flexible pipes. Examples are shown in Figure 11.19.

Flexible bedding

Pea gravel is the recommended bedding for flexible pipes.

Figure 11.20 shows a typical drain trench backfilled with the correct materials.

UPVC (plastic) pipes

UPVC pipes (Figure 11.21) are supplied with both ends plain for use with separate coupling. They are manufactured in longer lengths than clay pipes.

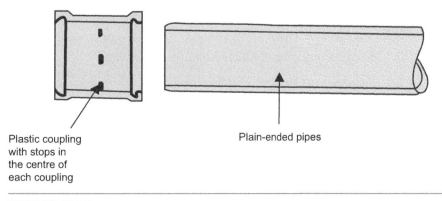

Plastic coupling with stops in the centre of each coupling

Plain-ended pipes

FIGURE 11.18
Plastic coupling

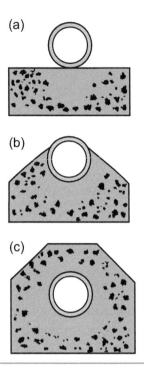

FIGURE 11.19
Rigid bedding: (a) bedded; (b) surrounded; (c) haunched

UPVC is flexible both as a pipe and at the joint. UPVC pipes must be correctly stored to prevent bending.

It does not have a crushing strength, so it must be bedded correctly to avoid continuous pressure deforming the pipe. Bedding for UPVC pipes is the same as for flexible pipes.

Pipes without sockets have to have separate couplings fitted (Figure 11.22).

Fittings

There are numerous fittings available in all shapes and sizes (Figure 11.23).

Some of the special fittings are:

- Double collars – These are used when repairing drains.
- Taper pipes – These are used when the internal diameter of the drain increases owing to the increase of flow of sewage.
- Junctions – These are used when tributary drains join the main drain without an inspection chamber being built.
- Channels – These can either be half or three-quarter section and are used in the base of inspection chambers to allow the drain to be opened for inspection and cleaning. They are also available in junctions and bends.

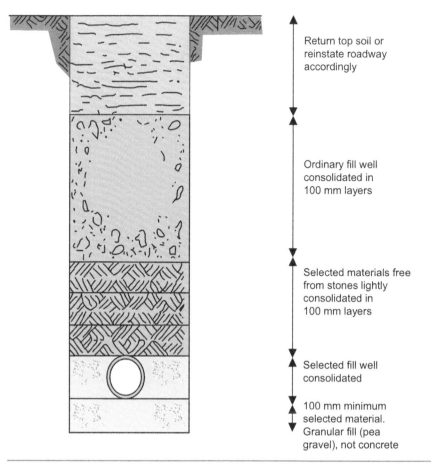

FIGURE 11.20
Backfill to drain trench

FIGURE 11.21
UPVC pipe

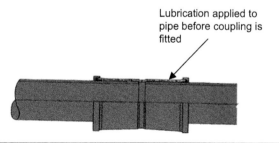

FIGURE 11.22
Coupling to pipes without collars

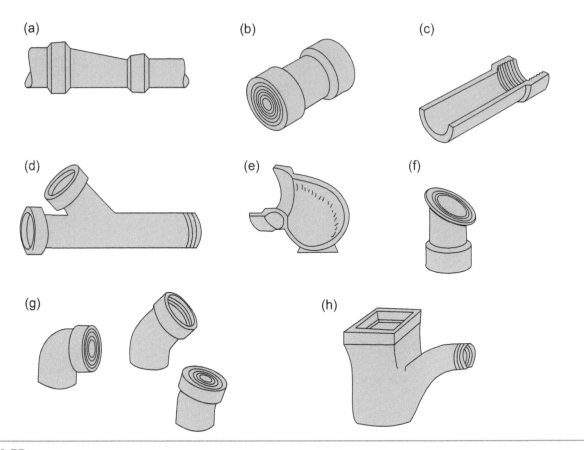

FIGURE 11.23
Various fittings: (a) taper; (b) double collar; (c) channel; (d) junction; (e) channel; (f) saddle; (g) bends; (h) gully

- Three-quarter channels – These are used to connect tributary drains to a main drain when the junction is being made within the inspection chamber. They are either right or left handed.

- Saddles – These are required when a new drain is to be connected to an existing one. A hole is carefully cut into the existing drain.

- When the hole is complete the saddle is bedded onto the existing pipe and allowed to set. The new drain is then connected to the saddle to allow the new drain to join the existing one.

- Bends – These are usually described by the degree of angle, such as 45 or 90 degrees. Very small quadrant bends are called knuckle bends and are used when connecting to a toilet or a rain water fall pipe.

- Gullies – Gullies are used to provide a water seal which prevents the foul smells and gases that build up in a sewer from escaping into the atmosphere.

Laying drains

As stated before, all drainage work must be carried out in accordance with the current Building Regulations.

Approved Document H2 lays down an acceptable level of performance necessary to reduce the risk of danger to the health and safety of people in the building.

1.   Convey the flow of foul water to a foul outfall (public sewer).

2.   Minimize the risk of blockage or leakage.

3.   Prevent foul air from the drainage system entering the building.

4.   Be ventilated.

5.   Be accessible for cleaning blockages.

Drainage is primarily divided into two groups:

- surface water

- foul water.

Surface water is water from roofs and the surrounding ground, whereas foul water is water which is contaminated by soil, waste and trade effluent.

Surface water does not require treatment and can be discharged into a local watercourse.

Any drainage system must convey all surface water and liquid sewage away from the building in the most speedy and efficient way, possibly to the sewer, without the risk of nuisance or danger to health and safety.

When designing any system the following principles must be observed:

- Provide adequate access points.

- Keep pipework as straight as possible between access points. An access point should be provided for all bends over 45 degrees.

- Ensure that all pipework is adequately supported.

- Ensure that the pipe is laid to a self-cleansing gradient.

- The whole system must be watertight, including inspection covers.

- Drains should not run under buildings, unless this is unavoidable or in so doing they would considerably shorten the route of the pipework.

Gradients

A drain should be designed with sufficient gradient to ensure that the speed of the liquid within the drain is such that it becomes self-cleansing.

A rule-of-thumb method for calculating the fall in a 100 or 150 mm diameter drain uses the following gradients:

- A 100 mm drain may have a fall of 1 in 40.

- A 150 mm drain may have a fall of 1 in 60.

- A 225 mm drain may have a fall of 1 in 90.

When levels are stated for drainage they will always refer to the invert level. It is vital that the gradient is consistent throughout the full length of a section of drain.

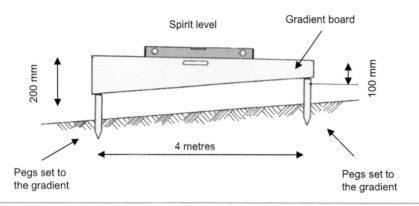

FIGURE 11.24
Setting out a 1 in 4 gradient

Changes of gradient or alignment must always occur within the confines of an inspection chamber.

A small drain trench can be set out with lines between the inspection chamber and the gully.

Method 1

A tapered straight edge (gradient board) could be used to check the correct gradient (Figure 11.24). Therefore, 4000 mm/100 mm = 1 in 40.

The pegs represent the level of the granular fill or concrete bed on which the drain will be laid.

Method 2: Boning rods

Boning rods (Figure 11.25) are made of timber in sets of three, which must be identical in size.

Procedure:

1. Peg 1 is driven into the required level.

2. Peg 2 is then driven in 1 m along the trench and is given the required fall, i.e. 1:40 = 25 mm in 1 m.

3. A boning rod is placed on both pegs 1 and 2.

4. Further pegs can now be driven in at any distance apart until the height of the peg plus the third boning rod is in line with the sight line of the other two boning rods.

Method 3: Sight rails

This method is similar to the boning rod method, but this time the fixed points are at either end of the drain run (Figure 11.26).

The site rails are set out a maximum 15 m apart along the proposed drain trench.

As the excavator digs the trench the banksman checks the depth using the traveller.

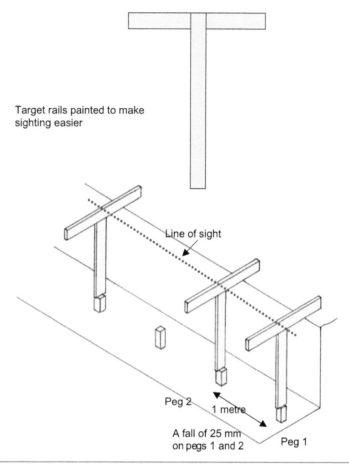

Target rails painted to make sighting easier

Line of sight

Peg 2

1 metre

A fall of 25 mm on pegs 1 and 2

Peg 1

FIGURE 11.25
Use of boning rods

The traveller acts in the same way as the third boning rod, but now extra length has been added to set out the bottom of the trench.

As excavation takes place the traveller moves down the trench establishing the correct fall of the trench bottom, which will be parallel to the line of sight.

When the trench has been excavated to the required gradient the bottom should be 'bottomed-up' to an even surface and uniform gradient.

The bottom of the trench is usually 100 mm deeper than required to allow the trench to be filled with 100 mm of suitable bedding material to receive the flexible drains.

Once the trench bottom is established the traveller can be shortened to invert depth and used to lay the drain pipes.

Access points

Access points are provided for clearing blockages from any drain runs that cannot be reached by any other means and can be either:

- inspection chambers
- rodding eyes.

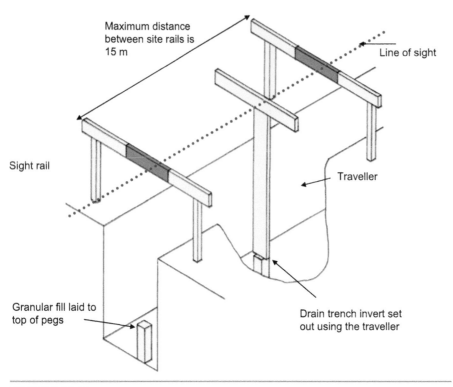

FIGURE 11.26
Site rails

All drains and sewers should have such means of access as may be necessary for inspection and cleaning. These should be placed:

- where there is a change in direction
- where there is a change in gradient
- at the highest point of the drain
- on a drain within 12.5 m of a junction of another drain or sewer unless an inspection chamber is at the junction.

No part of a drain should be more than 45 m away from an inspection chamber.

Inspection chambers can be constructed from brick, concrete and plastic.

Brick inspection chambers

Brickwork in inspection chambers is below ground and out of sight, but it must still be built correctly as the inspection chamber will have to be tested against water leaking out.

This section only provides a basic guide and operations could change around the country.

Inspection chambers are required for access to the drainage and sewerage systems. Brick chambers could also be required for access to other services.

The basic requirement of an inspection chamber is that it should be large enough to allow a person entrance and space to be able to work.

There may be branch drains entering the inspection chamber, so it should be designed according to the number of branches.

It has to be watertight and resist all possible loads from above and externally. The recommended minimum internal sizes are:

- 450 mm long × 450 mm wide when the inspection chamber is not more than 1 m deep
- 1200 mm long × 750 mm wide if it is more than 1 m deep

When inspection chambers have branches a minimum of 300 mm should be allowed for each 100 mm and 150 mm branch on the side with the most branches.

It is also advisable to allow 600 mm at the outlet end of the inspection chamber to allow for further connections to the inspection chamber and for easy rodding. The minimum width should be 450 mm or an extra 300 mm if there is a branch.

The inspection chamber should be constructed on a concrete base, normally 150 mm thick and built with one-brick thick walls in cement mortar not weaker than 3:1. The concrete base should be at least 150 mm thick.

Some authorities may specify class B clay engineering bricks (Figure 11.10), and sulphate-resisting cement if the ground contains sulphates.

The inspection chamber is normally built in English bond, but if there is high water pressure from outside water bond may be specified.

If the inspection chamber is over 750 mm deep, step irons should be built in every fourth course beneath the opening. They should be staggered, 300 mm apart, to allow easy access.

The cover frame should be 450 × 450 mm for inspection chambers less than 1 m deep and 600 × 600 mm for those more than 1 m deep

Construction of inspection chambers

The main channel pipe is laid when the main run of pipes is being laid by the drain layers. Any connecting branches are also laid and completed.

The concrete base is then laid in position around the channel pipes and allowed to set.

The brickwork is set out for the inspection chamber according to the number of branches required. Approximately six courses should be built, taking care to keep the inspection chamber plumb, level and square on the inside.

Any pipes through the walls should be protected from any pressure by turning small arches over those of 150 mm and above.

The benching should be completed at this level, when it is easier. Bricks could be used to form the basic structure of the benching and then finished smoothly with a stiff concrete. The gradient should be approximately 1:12 to prevent any waste from congregating on the benching.

Once the benching has hardened the remainder of the brickwork can be laid. If the inspection chamber is over 750 mm deep, step irons should be built in to assist in access to the chamber.

The inspection chamber is usually roofed over with an in situ reinforced concrete slab with an access hole. The brick shaft can be extended until the correct height is reached and the inspection cover fitted. Smaller inspection chambers may have the brickwork corbelled inwards to receive the inspection cover, instead of a concrete slab.

Brick inspection chambers have been superseded on many construction sites by precast concrete and plastic inspection chambers, which are circular on plan. The bricklayer may still be required to construct the small chamber on top of the concrete inspection chamber to receive the inspection cover.

Brickwork inspection chambers less than 1 m deep can be built as shown in Figure 11.27. Those more than 1 m deep can be built as shown in Figure 11.28.

Figure 11.29 shows details of arches, benching and corbelling to inspection chambers.

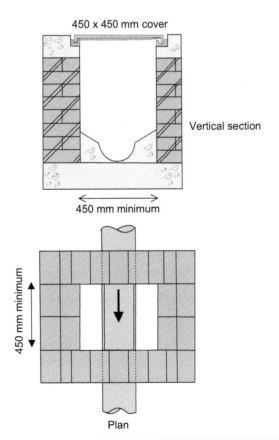

FIGURE 11.27
Small inspection chamber

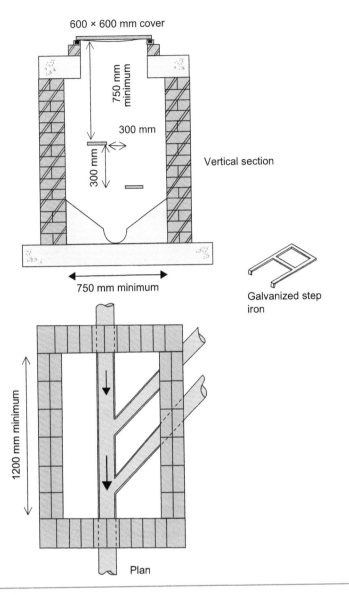

FIGURE 11.28
Large inspection chamber

Water bond is used to provide a watertight inspection chamber. Stretcher bond is used, but each leaf is built with bed and cross-joints staggered (Figure 11.30). Cement rendering is used between each leaf for added waterproofing.

Rodding eye

A rodding eye is used in place of an inspection chamber to enable the drain to be cleaned (Figure 11.31).

Plastic inspection chambers

The base unit is available with several inlets, all of which are blanked off. Those required for the design will need to be knocked out.

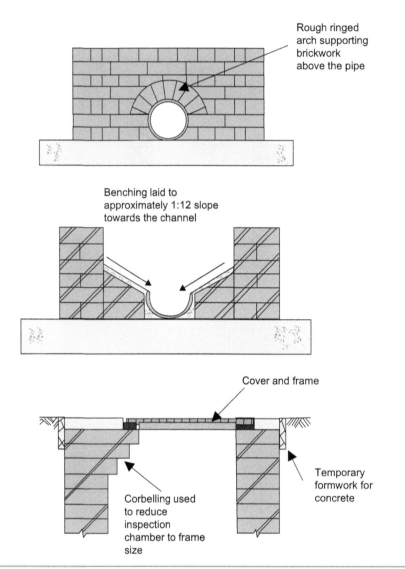

Rough ringed
arch supporting
brickwork
above the pipe

Benching laid to
approximately 1:12 slope
towards the channel

Cover and frame

Corbelling used
to reduce
inspection
chamber to frame
size

Temporary
formwork for
concrete

FIGURE 11.29
Inspection chamber details

The base is bedded on a shingle base or 150 mm wet concrete to ensure a uniform base. Raising pieces, each 650 mm deep, are added until the required height is reached. They can be cut with a saw to reach the required depth. A cover and frame are fitted to the top when the required height has been reached.

Precast concrete inspection chambers

Once the drain run has been completed, the channel pipes have been set and the branch connections made if required, the concrete base can be placed as for brick inspection chambers. More concrete should be placed to form the required benching to form a level base to receive the first chamber section. The chamber sections are either rectangular or circular on plan.

The first chamber section should be bedded onto the concrete base. Further chamber sections are bedded until the correct height is reached. A cover and

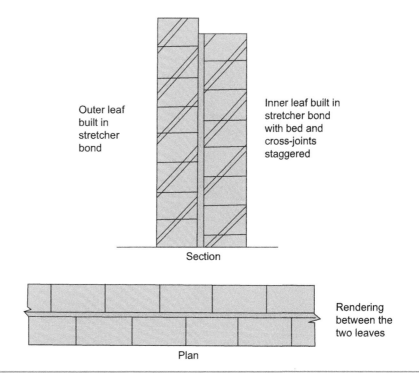

FIGURE 11.30
Water bond

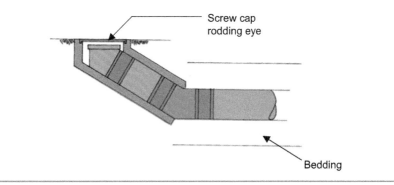

FIGURE 11.31
Rodding eye

surround are then positioned; these are available in various sizes and strengths. For deeper chambers sections are available with step irons already fitted.

Plastic inspection chambers

There are various designs of plastic inspections chambers as shown in Figure 11.32.

Precast concrete manholes

There are also many designs made from precast concrete which are shown in Figure 11.33.

(a)   (b)

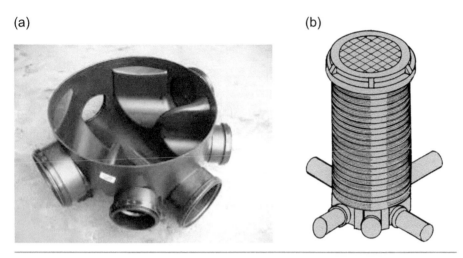

FIGURE 11.32
Plastic manhole base (a) and completed plastic manhole (b)

(a)   (b)

FIGURE 11.33
Round and square precast concrete manholes

Drain testing

After the pipe run has been completed it should be tested. All drains have to be tested before and after backfilling.

There are three main methods of testing for watertightness:

- water
- air
- smoke.

The most common method is the air test, as the water test could cause problems with the removal of the water after the test.

Equipment and procedure for carrying out the air test

The following equipment is required (Figure 11.34):

- Expanding drain plug – These are available in sizes from 50 to 300 mm. They are placed into the drain and tightened.

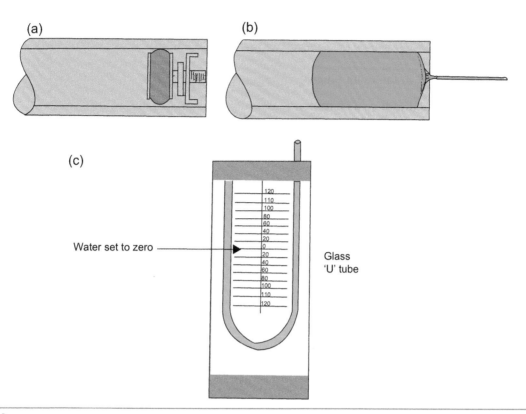

FIGURE 11.34
Drain testing equipment: (a) expanding drain plug; (b) inflatable stopper; (c) U gauge

- Inflatable stopper – These are canvas-covered rubber air bags which are placed into the drain and then inflated from above.
- U gauge – The U gauge is connected to the stopper at the highest end.

Inspect the drain system for possible damage during or after laying.

Check the condition of and fit the drain plugs to all ends of the drain.

After the drain has been sealed with drain plugs, air is pumped into the drain to approximately 0.021 N/mm^2 (using a hand pump or bellows).

The U gauge is connected to the highest stopper: 100 mm of pressure should be indicated on the U gauge connected to the system. The pressure should not drop by more than 25 mm in five minutes.

Procedure for carrying out the smoke test

Inspect the drain system for possible damage during or after laying. Check the condition of and fit drain plugs at the highest ends.

Before the lowest drain plug is fitted a smoke bomb is lit and placed inside the drain before fitting the last drain plug.

Any leaks will be spotted by smoke coming from the joints in the drain system.

Procedure for carrying out the water test

As with all methods, inspect the drain system for possible damage during or after laying.

Check the condition of and fit the drain plugs to lower ends of the drain.

If rigid joints have been used with cement mortar joints, allow 24 hours to dry before testing.

With the water test 1.5 m of head is required. This can be achieved by attaching a pipe to the highest point of the drain (Figure 11.35).

Fix the stopper to the lowest points of the drain and fill the drain with water to the top of the header pipe. Allow to settle and then check for leaks. Some water may be lost in absorption – top up the header pipe as required.

The loss of water is now checked over a period of 30 minutes and loss should not exceed 1 litre per hour, per linear metre, per metre of nominal internal diameter.

Example 6

Given a drain run, 30 m long in 300 mm diameter pipes, the permitted allowable loss per 10 minutes is:

$$\frac{1 \times 10 \times 30}{60} \times \frac{300}{1000} = 1.5 \text{ litres}$$

Testing inspection chambers

Inspection chambers are generally tested by fitting an inflatable stopper in the outlet of the inspection chamber, so the air can be removed at ground level. All the remaining inlets and outlets are also stopped up.

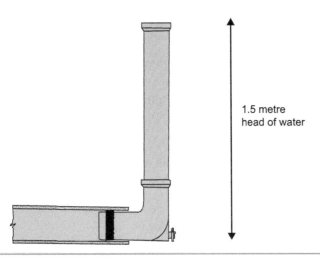

1.5 metre head of water

FIGURE 11.35
Calculation of loss of water

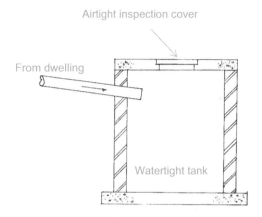

FIGURE 11.36
Cesspool construction

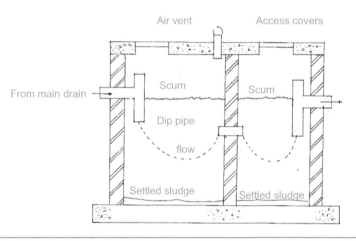

FIGURE 11.37
Septic tank

The chamber should then be filled with water and left to stand for eight hours. Topping up can be carried out as necessary.

The acceptable criterion is that no water should be observed issuing from the outside of the chamber.

Backfilling

The drain trench can then be backfilled in 100 mm layers for 300 mm. Subsequent backfilling can be carried out using excavated material in layers not exceeding 300 mm.

Disposal of sewage in rural areas

Properties built in the countryside are not always able to be connected to the public sewers.

Waste water and foul water must therefore be discharged in a different way than in urban areas.

The two main methods are by using a cesspool or a septic tank.

Cesspools

A cesspool is a sealed, watertight underground tank that captures wastewater and sewage and simply collects it, without processing or treating it in any way.

Cesspools will usually be discreetly buried underground with a single manhole cover for access.

These need emptying by a tanker that sucks out the effluence at regular intervals by the Local Authority waste collection team.

Cesspools are recommended where septic tanks are unacceptable to an authority or where impermeable ground conditions exist.

Building Regulations require a minimum capacity for a single dwelling should be at least 18 cubic metres.

Cesspit

Cesspits were normally constructed by brickwork in circular or rectangular design and could be mistaken to look like a well in appearance. In the early days 1900-1930's cesspits were common especially in period properties not served by mains drainage.

They would have been constructed by hand to create chamber normally no deeper than 2 meters by 1 meter wide. They were constructed with brick to create the cesspit. The brickwork was positioned 50 mm apart to allow gap for the untreated effluent to soakaway through the wall and the dirt base of the cesspit. Visual appearance of the cesspits would have thick concrete or natural stone slab positioned over the access point to act as inspection lid.

Cesspits are very basic example of off mains drainage system with the tank and soakaway combined to make all in one design for disposal of wastewater.

The logic was there at the early stages for off main drainage but the designed was floored with many problems occurring.

Under no circumstances can cesspits be used as off mains drainage system as there is high risk of pollution.

From 1936 the Public Health Act prohibited the use of cesspits and other methods off drainage system would have to be used such as cesspools or septic tanks.

Septic tanks

There are areas when connection to the main sewer is not possible and other options have to be employed.

FIGURE 11.38
Prefabricated septic tank

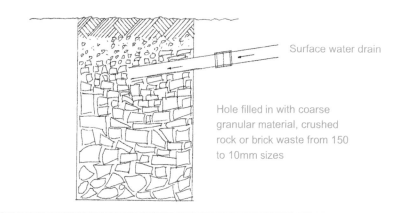

Surface water drain

Hole filled in with coarse
granular material, crushed
rock or brick waste from 150
to 10mm sizes

FIGURE 11.39
Section through a soakaway

A septic tank is an underground chamber made of brickwork, concrete, fiberglass, or plastic through which domestic wastewater flows for basic treatment on your own land as shown in Figure 11.37.

A septic tank is similar to a complete waste disposal system in which the sewage is broken down into sludge by bacterial action before the water is discharged into a local watercourse or by percolation into the surrounding subsoil.

Settling and anaerobic processes reduce solids and organics, but the treatment efficiency is only moderate.

There are numerous designs of prefabricated septic tanks in glass reinforced polyester or fibreglass which are lowered into an excavated hole and connected to the house drains. These are very efficient and save time and money in constructing a concrete and brick chamber. See Figure 11.38.

Soakaways

Soakaways can only be used on top water drainage.

They are used when it is difficult to connect the outlet to the main top water drain, therefor saving time and expense.

A soakaway consists of a large hole dug in the ground to a depth and size according to the amount of top water anticipated.

The soakaway must be at least 5 metres from the building and 2.5 metres from your neighbours' fence. The pipe flowing to the soakaway should be 100mm underground waste ideally, but the Building Inspector may settle for 75mm, and it should be laid to a fall of 1:40.

The hole is filled with porous material and the drain is allowed to enter this porous material. The top of the filling is then sealed off with heavy duty landscaping fabric material that will not allow the top soil to penetrate the porous filling.

Multiple-choice questions

Self-assessment

This section of the book is designed to allow you to check your level of knowledge. The section consists of revision questions for this chapter. The questions are all multiple choice and have four possible answers. The answers are to be found at the end of the book.

The main type of multiple-choice question will be the four-option multiple-choice question. This will consist of a question or statement, known as the stem, followed by a choice of four different answers, called the responses. Only one of these responses is the correct answer; the others are incorrect and are known as distracters.

You should attempt to answer the questions by choosing either (a), (b), (c) or (d).

Example

The person employed by the local authority to ensure that the Building Regulations are observed is called the:

    (a) clerk of works

    (b) building control officer

    (c) council inspector

    (d) safety officer

The correct answer is the building control officer, and therefore (b) would be the correct response.

Domestic drainage

Question 1 Identify the following drain fitting:

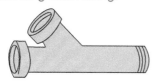

    (a) junction

    (b) saddle

    (c) double collar

    (d) bend

Question 2 The maximum distance between access points must not be greater than:

    (a) 35 m

    (b) 40 m

    (c) 45 m

    (d) 50 m

Question 3 Calculate the number of drain pipes required for a trench 13.5 m
long if 0.900 mm pipes are being used:

(a) 15

(b) 13

(c) 11

(d) 9

Question 4 Which drainage system conveys waste water from both surface water
and foul water in a single pipe to the sewage treatment works?

(a) separate system

(b) combined system

(c) partially separate system

(d) joint system

Question 5 What is the significance of a black band printed on a drain pipe?

(a) first class pipe

(b) can only be used in separate systems

(c) second class pipe

(d) can only be used in combined systems

Question 6 What is a joint caulked with tarred hemp and finished with cement
mortar known as?

(a) a flexible joint

(b) a solid joint

(c) a patent joint

(d) a rigid joint

Question 7 What is the recommended gradient for a 150 mm drain?

(a) 1 in 30

(b) 1 in 40

(c) 1 in 50

(d) 1 in 60

Question 8 Identify the equipment shown, used for setting out drain trenches:

(a) gradient boards

(b) boning rods

(c) profile boards

(d) site rails

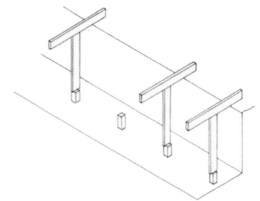

CHAPTER 12

Answers to Multiple-Choice Questions

Chapter 1. The Construction Industry (Community Buildings)

1 (a); 2 (c); 3 (b); 4 (d); 5 (b); 6 (c); 7 (c); 8 (a)

Chapter 2. Health and Safety in the Construction Industry

1 (c); 2 (c); 3 (c); 4 (b); 5 (c); 6 (b); 7 (c); 8 (d)

Chapter 3. Communication and Records

1 (a); 2 (b); 3 (a); 4 (c); 5 (a); 6 (b); 7 (c); 8 (b)

Chapter 4. Construction Technology

1 (d); 2 (a); 3 (b); 4 (b); 5 (a); 6 (b); 7 (d); 8 (c)

Chapter 5. Moving and Handling Resources

1 (b); 2 (c); 3 (d); 4 (a); 5 (a); 6 (c); 7 (d); 8 (d)

Chapter 6. Solid Walls and Piers

1 (b); 2 (d); 3 (a); 4 (c); 5 (a); 6 (d); 7 (a); 8 (c)

Chapter 7. Constructing Cavity Walls with Openings

1 (b); 2 (c); 3 (a); 4 (c); 5 (d); 6 (c); 7 (a); 8 (b)

Chapter 8. Cladding to Timber, Steel and Concrete Frames

1 (b); 2 (b); 3 (a); 4 (d); 5 (c); 6 (c); 7 (a); 8 (b)

Chapter 9. Thin Joint Masonry

1 (b); 2 (c); 3 (d); 4 (a); 5 (a); 6 (d); 7 (c); 8 (b)

Chapter 10. Bridging Openings with Lintels and Arches

1 (a); 2 (b); 3 (c); 4 (d); 5 (c); 6 (a); 7 (b); 8 (c)

Chapter 11. Domestic Drainage

1 (a); 2 (c); 3 (a); 4 (b); 5 (c); 6 (d); 7 (d); 8 (b)

Index

Abrasive wheels 212
Abutments 374
Accidents 20, 27
Aggregates 159
Air bricks 115, 275, 318
Anchor straps 292
Arch
    formers 381
    support 381
    terms 370
Arches 368–392
    arch terms 370
    axed 370
    centres 381–382
    classification 369
    collar joint 376
    discharging 380
    gauged 370
    geometry 372–377
    key brick 373
    relieving 370
    rise 372
    rough ring 370
    segmental 375
    semi-circular 372
    skewback 378
    soldier 369
    span 372
    types 369
    voussoirs 371
Atmospheric depth 109
Attached piers 241–242

Back actor 102
Banding 207
Bar charts 37
Bevelled closers 238
Blocks
    bearing 136
    clay blocks 137
    concrete 137
    setting out 278
    special blocks 138
    types 137
Block plans 388
Boat level 255
Bonding 208
    broken bond 227–232, 280
    dry bonding 214
    problems 226

principles 208
purpose 226
reverse bond 233
rules of 211–213
Bonds *see separate names*
Boning rods 408
Boundary wall 233, 249
Boxed materials 169
Brick cladding 319
    details 332
    eaves details 324
    fixing 323
    foundation 322
    moisture barrier 325
    support 322
    window head 324
    window sill 324
Brick on edge 252
Bricks
    actual size 222
    commons 132
    co-ordinating size 222
    engineering 132
    facings 132
    identification 173
    laying to the line 217
    paving 173
    purpose made 133–134
    quality 133
    refractory 133
    slips 325
    storage 187
    transporting 207
    types 133–134, 173
    and see separate names
Bricks and mortar quantities 57–65
Bridging openings 360
Broken bond 227–232
BS 1192 36
Builders 2
Building defects 141
    chemical attack 143
    dry rot 144
    fungal attack 143
    general 141
    insect attack 144
    moisture 143
    protection and prevention 143
    wet rot 144
Building line 197

Building regulations 265
Building sand 159
Buildings
    communal 4–6
    communications 4–6
    dwelling units 4–6
    industrial 4–6
    recreational 4–6
Bulk material 156

Cabins 30
Calculations
    blockwork 62–64
    brickwork 57–62
    cavity walls 68
    centre line 71–75
    circles 81–84
    cost 56
    cubic 55
    decimals 56
    division of commons to facings 79
    drainage 393–394
    linear 55
    metric system 57
    mortar 58, 65
    number 54
    openings 67
    percentage 78
    perimeters 69–71
    superficial 55
    time 56
    triangles 81
    types 54
    waste 80
Cappings 250–251
Cavity wall 264
    battens 282
    cleaning 283
    construction of 278
    closing at eaves 292
    definition 264
    DPC 286
    DPC trays 287
    function 265–268
    insulation 277
    jambs 285
    lintels 285
    loads 267
    openings 285
    purpose 264
    quoins 274
    reveals 279, 286
    reverse bond 281
    setting out bonding 279
    setting out materials 278
    sills 287
    thresholds 288
    ties 274
    vertical DPC 286
    weep holes 268
Cavity walling
    advantages 269
    definitions 264
    disadvantages 269
    functions 267

insulation 277
    purpose 264
    regulations 265
    requirements 268
CDM regulations 18
Cement
    Portland 160
    storage 151
Centre line method 71–75
Cesspit 420
Cesspools 419
Circuit breakers 205
Cladding
    definition 312
    fixings 317
    functions 312
    types 312
Classes 6
    business 7
    HMOs 8
    hotels 7
    leisure 8
    residential 8
    shops 7
    sui generis 8
Close poling boards 103
Collar joint 216–218
Combined system 400
Communication
    barriers 89
    external 91
    internal 91
    methods 85
    negative 89
    notes 89
    oral 86
    positive 89
    telephone 86
    written 86
Community buildings 4–6
Companies 2
Company policy 38
Compound 26
Compression joints 326–327
Concrete corbel 303
Concrete frames 323–331
    brick panels 330
    brick slips 325, 328
    compression joints 325, 327, 331
    definitions 323
    DPC trays 325–326
    fixings 330
    movement joints 330
    support 325, 331
Conduction 265
Contract of employment 43
Construction industry 2
Construction process 3
Containerised materials
    storage 192
    types 166
Contractors 2
Copings 250–251
Corbelling 301
Coring holes 283, 285

Corner blocks 216
COSHH 18,
Cross joints 218
Cutting to rake 242–245
Cutting up a gable 302–307

Damp proof course/membrane
    see DPC/DPM
Dampness penetration 143
Datum peg 100
Daywork sheets 40–41
Deadman 243
Detached piers 233
Development plan 90
Discharging arch 380
Dividers 373
Double returns 223, 225
Dovetail slot 330
DPC
    bedding 216, 251
    cavity trays 287
    engineering bricks 216
    location 115
    storage 167
    vertical 286
DPM 110, 322
Drainage 386–410
    access points 410
    backfill 405
    benching 414
    calculations 393–394
    fittings 404–406
    gradients 407
    laying 407
    plans 386–391
    safe working 394
    schedules 391–392
    systems 398
    testing 416–418
Drain pipes
    joints 402–403
    storage 187
    types 175, 396, 402
Drawings 36
    assembly 402
    basic methods 48
    elevations 52
    layout of drawing 52
    plans 52
    producing working drawings 42
    projections 49–50
    range 390
    register 44
    scales 46–47
    schedules 392
    sections 52
Dry bonding 214, 279
Dry rot 143
Dutch bond 226

Eaves detail 128
Electricity
    circuit breakers 205
    connectors 204

on site 203
    plugs 205
Electricity at Work Regulations 18
Emergencies 27
Employees 2
Engineering bricks 216
English bond 217, 220–222
English garden wall bond 227
Estimating 84
Excavations 101
External envelope 120
Extrados 370
Eye lines 252

Feather edge coping 251
Fire precautions regulations 18,
Fixing gates 247–249
Fixings 167
Fixings for slip bricks 328
Flammable liquids storage 191
Flat packs 168
Flat roofs 127–129
Flemish bond 224
Flemish garden wall bond 227
Floors
    boarding 116
    finishes 116
    floating 117
    hollow ground floors 115
    precast beam and pot 116
    screeds 116
    solid ground floors 113
Folding wedges 373
Foul water 398
Formwork 365
Foundations
    atmospheric depth 109
    design 109
    general 108
    level pegs 104
    loads 108
    minimum width 109
    narrow strip 116
    pad 111
    profiles 104
    rafts 112
    short bored piles 112
    standard strip 109
    types 109
    wide strip 110
Fragile materials 157
Frames 293–297
    checking fixing 294
    fixing door frames 295
    fixing window frames 295
    protection 297
    temporary 296
    types of fixings 295
Frontage line 97
Fully filled cavity insulation 277

Gable ends 302
Gallows bracket 303
Garden wall bonds

English 226
Flemish 227
Gate pillars 245–248
Gauge
 horizontal 254
 rods 106
 vertical 106
Geometry of arches 372–377
Gradient boards 407
Ground floors
 solid concrete 113
 suspended timber 115
Ground works 101
Guide rafters 302
Gun templet 378

Half bond 209
Handling materials 178
Hardwoods 162
Hazardous material 156–157
Hazards 21
Header bond 221
Health and hygiene 30
Health and Safety at Work Act 18–20
Heat loss 265
Heritage 10
Honeycomb sleeper wall 115

Information
 interpreting 44, 386
 sources 36, 168, 200
 storing 44
Injuries 178
Inspection chambers 396–416
 benching 414
 construction of 411
 covers 414
 channel 414
 requirements 412
 testing 418
 water bond 415
Insulation
 brick and block walls 120
 fully filled 277
 partial filled 277
 timber frame walls 120
 types 317
Intrados 370
Ironmongery 167
Isolated piers 237–240
Isometric projection 50

Jigs 296

Kerbs
 types 175
 storage 188
Key brick 370

Legislation 9, 19
Levels
 transferring levels 106
Lifting gear 179
 mechanical 180–181
Lime 160

Lime leaching 251
Lintels 360–368
 bearings 368
 bedding 368
 boxed steel 366
 cast in situ 363
 casting on site 363
 channelled steel 366
 corrugated steel 366
 definition 362
 formwork 365
 pre-cast 362
 pre-stressed 364
 reinforced brick 367, 369
 reinforced concrete 362–363
 reinforcing bars 365
 steel sections 366–367
 stresses 363
Location plan 389
LOLER 18

Maintenance work 2
Manual handling regs 18
Manufacturer's technical information 39, 170, 340, 360
Masonry saws
 bench 202
Mastic sealant 258
Material records 145
 checking deliveries 145
 controlling the use 146
 ordering materials 146
 planning deliveries 146
 transfers 150
Materials
 aggregates 130, 159
 bagged 160
 boards 162
 boxed 167
 blocks 174
 bricks 173
 bulk 156
 cement 130
 characteristics 139, 156
 concrete 131
 containerised 166
 drainage 175
 fragile 157
 handling 179
 hazardous 156–157
 insulation 163
 ironmongery 167
 lime 131, 160
 miscellaneous 158
 moving 270
 paints 166
 perishable 157
 plasterboards 164
 properties 130
 reinforcement 164
 rolled 167
 sheet 40, 162
 stock rotation 151
 stone 137
 storing materials 148

timber 160
   timber products 160–164
Mission statement 38
Modular construction 335
Mortar 131
Movement joint 257–258
Moving materials 270
Multiple-choice answers 425

Narrow strip foundation 116
New work 2
Noise at Work Act 20

Oblique projection 51
Office facilities 31
Openings
   history 360
   lintel schedule 360
Ordinary Portland cement 160
Orthographic projection 49
Oversailing courses 257
Oversite concrete 115

Pad foundations 111
Parapet walls 249
Partial fill cavity insulation 277
Partially separate system 401
Paving slabs
   types 175
   storage 188
Perishable materials 157
Perpends 218–219
Personal hygiene 30
Piers
   attached 241–244
   bonding 237–238
   construction 235–239
   detached 233
   foundations 236
   problems 234
Pinch rod 297
Pistol bricks 326, 328
Plasterboards
   types 164
   storage 190
Plywoods 162
Poling boards 103
Portland cement 160
PPE regulations 18
   types 21–22
Prefabricated units 336
Preservation 12
Profile boards 99
Programmes of work 37
Proprietary wall connectors 299
PUWER 18

Quantities of materials 53
Quarter bond 209
Queen closer 222
Quoins 215, 223–224, 274, 281, 300

Racking back 215
Raft foundations 112
Raising quoins 214–215

Raking cutting 242–245, 302–307
Ramps 241
Ranging in 215
Ranging lines 104
Recording materials 41
Refurbishment 3–10
Reinforcement 164, 256–258
Reinforcing mesh 164, 298, 318
Renovation 13
Requisition sheets 41
Restoration 3–10
Reveal bricks 280
Reverse bond 233
RIDDOR 18
Ridge detail 128
Right angle
   3:4:5 method 100
Rise 370
Risk assessments 29
Rodding eye 415
Rolled materials
   types 167
   storage 192
Roof truss 322
Roofing tiles
   types 176
   storage 189
Roofs
   basic forms 126
   components 127
   flat 127–129
   terminology 127
Rules of bonding 211–213
Running laps in DPC 315

Saddle-back capping and
   coping 251
Safety
   legislation 18
   signs 21–25
Sand
   building 159
   courses 107, 247
Scaffolding
   access 271
   toe boards 272
   guard rails 272
Scales 47–49
Schedule 37–39
Screeds 116
Sectional bonding 219–220
Security 25
Separate system 400
Septic tank 419
Service posts 395
Services 107
Setting out 96
   a building 97
   accuracy 97
   checking for accuracy 98
   design 98
   face brickwork 210
   horizontal measurement 98
   marking out 101
   materials 206

profile boards 99
right angle 100
Sheet materials 189
Short bored piles 112
Sills 289
SIPs 334
Site accommodation 32
Site compound 26, 183
Site diary 40–42
Site plan 45, 388
Site rails 410
Site security 26
Site storage 183
Skewback 378
Skips 194
Sleeper walls 115
Sliding bevel 306
Slip bricks 325
Soakaway 421
Softwoods 160
Soldier
    arch 368
    course 255
Solid ground floor 113
Span 370
Special shape bricks 134
Specification 37–39, 46
Spot boards 206
Springer brick 377
Springing line 372
Standard strip foundation 109
Steel frames 331–333
    column cladding 323
    definition 331
    foundation 332
    panel walls 332
Steel pegs 104
Step irons 397, 413
Stopped end 223
Storey rod 106
Straight joints 209
Stretcher bond 216, 227, 269
Striking point 370
String courses 254
Stone 137
Surface water 398

Tee-junctions 223, 224
Templet
    gun 378
Testing drains
    air test 417
    smoke test 418
    water test 417
Thermal insulation 277
Three:four:five method 100
Threshold 268
Thin joint masonry 340–356
    application 345
    benefits 346
    block types 341–342
    broken bond 349
    definition 340
    fixings 353
    junction walls 350–351

laying 347–352
lintels 356
mortar 342–343
movement 354
plumbing 348
profiles 352
quoin 347, 352
reinforcement 355
reverse bond 350
stability 352
tools and equipment 343–344
Tiled
    creasing 252
    sill 288
Timber frames 124–126
    balloon frame 320
    definitions 319
    platform frame 321
    stick system 321
Timber storage 189
Timbering to trenches 103
Time sheets 40
Tingle brick 216
Tingle plate 217
Toothings 216, 299
Traveller 409
Trench fill foundations 116
Turning piece 378

Underfloor ventilation 115, 276
Unit of bond 220
U-values 267

Vapour/ moisture barrier 322–325
Variations 85
Vertical DPC 287
Vertical movement joints 259, 300, 329
Vibration 18,
Voussoirs 372

Waste
    minimising 152
    disposal from heights 194
Water bar 289
Water bond 415
Walls
    cavity 117
    external 126
    functions 122
    internal 122
    junctions 297–299
    load bearing 123
    non load bearing 123
    partitions 123
    timber frame 118
Wall plate 292
Wall ties
    position 275
    types 274
Weathering 251
Weep holes
    position 268, 291
Welfare facilities 28–30
Welsh arch 107
Wet rot 143

Wide strip foundation 110
Window frame storage 191
Work at Height Regulations 18
Work below ground level 101–105
  concrete foundation 105
  preparation of materials 105
  procedures 102

ranging lines 104
timbering 103
transferring levels 104
Working at heights 270
Working platforms 272
Working space 206
Workplace safety 21

Printed and bound by CPI Group (UK) Ltd, Croydon, CR0 4YY

24/10/2024

01778298-0016